Keratinocytes, as the main cellular component of the organism/environment interface, perform a vast range of functions in protection, secretion, sensation and self-repair by virtue of their great plasticity of form and development. Indeed, recent medical advances in laboratory culture of these cells for use as skin grafts in cases of severe burns or ulceration owe much of their success to this very plasticity.

This very practical handbook draws upon a wide range of international expertise making a collection of methods suitable for keratinocyte manipulation for use by new and old hands alike.

Together with its companion volume *The Keratinocyte Handbook* it provides a comprehensive dossier of the keratinocyte, which will prove an invaluable reference resource for all encountering keratinocytes in both laboratory and clinic, including dermatologists, molecular and cell biologists and biochemists.

Keratinocyte methods

Keratinocyte methods

Edited by

I RENE M. LEIGH and F IONA M. WATT

CAMBRIDGE
UNIVERSITY PRESS

Published by the Press Syndicate of the University of Cambridge
The Pitt Building, Trumpington Street, Cambridge CB2 1RP
40 West 20th Street, New York, NY 10011-4211, USA
10 Stamford Road, Oakleigh, Melbourne 3166, Australia

© Cambridge University Press 1994

First published 1994

Printed in Great Britain at the University Press, Cambridge

A catalogue record for this book is available from the British Library

Library of Congress cataloguing in publication data

Keratinocyte : methods / edited by Irene M. Leigh and Fiona M. Watt.
   p. cm.
Includes bibliographical references and index.
ISBN (invalid) 0-521-45103-6 (hc)
1. Keratinocytes—Laboratory manuals.   I. Leigh, Irene (Irene M.)
II. Watt. Fiona M.
[DNLM: 1. Keratinocytes—cytology.   2. Cells, Cultured,   WR 101
K395 1994]
QP88.5.K44   1994
611′.0181—dc20
DNLM/DLC
for Library of Congress   93-42614
                                               CIP

ISBN 0 521 45013 6 Keratinocyte methods
ISBN 0 521 43416 5 Keratinocyte handbook
ISBN 0 521 45878 1 Set of two volumes

PR

Contents

List of contributors

B. L. ALLEN-HOFFMANN
5638 Medical Sciences Centre, 1300 University Avenue, Madison, WI
53706, USA

V. BOUVARD
ICGEB, Unido, Padriciano 99, 1.34012 Trieste, Italy

J. CARROLL
Department of Oral Biology and Pathology, School of Dental
Medicine, State University of New York at Stony Brook, New York
11794-8702, USA

A. DALLEY
Experimental Dermatology Research Laboratories, London Hospital
Medical College, 56 Ashfield Street, London E1 1BL, UK

M. DE LUCA
Laboratorio di Differenziamento Cellulare, IST, Instituto Nazionale per
la Ricerca sul Cancro 16132, Genova, Italy

R. DOVER
Histopathology Unit, Royal College of Surgeons of England,
35–43 Lincoln's Inn Fields, London WC2A 3PX, UK

R. EADY
Department of Cell Pathology, St. John's Institute of Dermatology,
St Thomas's Hospital, Lambeth Palace Rd, London SE 17EH,
London, UK

C. FISHER
Cancer and Infectious Diseases Research, Upjohn Company,
Kalamazoo, MI 49001-0199, USA

N. FUSENIG
Division of Differentiation & Carcinogenesis *in Vitro*, Institute of
Biochemistry, German Cancer Research Centre, Im Neuenheimer Feld
280, D-6900 Heidelberg, Germany

J. GARLICK
Department of Oral Biology and Pathology, School of Dental
Medicine, State University of New York at Stony Brook, New York
11794-8702, USA

H. GREEN
Department of Cellular and Molecular Physiology, Harvard Medical
School, 25, Shattuck Street, Boston, MA 02115, USA

J. HANSBROUGH
Department of Surgery, University of California, San Diego Medical
Centre, 200 West Arbour Drive, San Diego CA 92103, USA

H. HENNINGS
Laboratory of Cellular Carcinogenesis and Tumor Promotion, National
Cancer Institute, Bethesda, MD 20892, USA

N. HOTCHIN
Keratinocyte Laboratory, ICRF, PO Box 123, Lincoln's Inn Fields,
London WC2A 3PX, UK

G. HOWELLS
Department of Oral Pathology, London Hospital Medical College,
London E1 1BB, UK

A. ISHIDA-YAMAMOTO
Department of Dermatology, Asahikawa Medical College, Nishikagura
4-5-3-11, Asahikawa, 078 Japan

P. H. JONES
Keratinocyte Laboratory, ICRF, PO Box 123, Lincoln's Inn Fields,
London WC2A 3PX, UK

T. KEALEY
Department of Clinical Biochemistry, University of Cambridge,
Addenbrooke's Hospital, Hills Road, Cambridge CB2 2QR, UK

I. M. LEIGH
Royal London Hospital Medical College, 56, Ashfield Street, London
E1 2BL, UK

J. McGRATH
Department of Cell Pathology, St John's Institute of Dermatology,
St Thomas's Hospital, Lambeth Palace Road, London SE1 7EH, UK

I. A. McKAY
Experimental Dermatology Research Laboratories, London Hospital
Medical College, 56 Ashfield Street, London E1 2BL, UK

R. MITRA
Department of Pathology, University of Michigan Medical Center, Ann
Arbor, Michigan 48109-0602, USA

R. J. MORRIS
Lankenau Medical Research Center, 100 Lancaster Ave. West of City
Line, Wynnewood, Pennsylvania 19096, USA

H. A. NAVSARIA
London Hospital Medical College, 56 Ashfield Street, London
E1 2BL, UK

B. NICKOLOFF
Department of Pathology, University of Michigan Medical Center, Ann
Arbor, Michigan 48109-0602, USA

W. OTTO
ICRF, PO Box 123, Lincoln's Inn Fields, London WC2A 3PX, UK

N. PARENTEAU
Organogenesis Inc. 150 Dan Road, Canton, Ma 02021, USA

J. PAVLOVITCH
CNRS URA 583, Hôpital des Enfants-Malades, Tour Lavoisier 6e
étage, 149 Rue de Sevres, 75743 Paris Cedex, France

M. PHILPOTT
Department of Clinical Biochemistry, University of Cambridge, Addenbrooke's Hospital, Hills Road, Cambridge CB2 2QR, UK

B. C. POWELL
Department of Biochemistry, North Terrace University of Adelaide, Adelaide, South Australia 5005

R. H. RICE
Department of Environmental Toxicology, University of California at Davis, Davis, CA 95616, USA

M. RIZK-RABIN
CNRS URA 583, Hôpital des Enfants-Malades, Tour Lavoisier 6e étage, 149 Rue de Sevres, 75743 Paris Cedex, France

G. E. ROGERS
Department of Biochemistry, North Terrace, University of Adelaide, Adelaide, South Australia 5005

E. L. RUGG
Cancer Research Campaign Laboratory Department of Anatomy and Physiology, University of Dundee, Dundee DD1 4HN, UK

C. SEXTON
Wellcome Research Laboratories, Langley Coast, South Eden Park Road, Beckenham, Kent BR3 3BS, UK

N. SHEIBANI
5638 Medical Sciences Centre, 1300 University Avenue, Madison, WI 53706, USA

J. TEUMER
Department of Cellular and Molecular Physiology, Harvard Medical School, 25, Shattuck Street, Boston, MA 02115, USA

F. M. WATT
Keratinocyte Laboratory, ICRF, PO Box 123, Lincoln's Inn Fields, London WC2A 3PX, UK

K. ZEZULAK
Department of Cellular and Molecular Physiology, Harvard Medical School, 25, Shattuck Street, Boston, MA 02115, USA

Preface

We are frequently asked for advice about how to culture keratinocytes and how to analyze their growth and differentiation. While preparing *The Keratinocyte Handbook* we therefore decided to compile a set of methods from our own laboratories and from our colleagues that readers might find useful. The resulting volume can be perused along with the main Handbook, or can be read independently. In it, you will find a combination of detailed protocols, recipes for culture medium and comments on the potential problems and pitfalls of applying widely used experimental techniques, such as flow cytometry and *in situ* hybridization, to the study of keratinocytes. We assume that workers who use this book will already be familiar with procedures for the safe handling and disposal of human cells and tissues, of retroviruses, oncogenic DNA, chemicals and radio-isotopes, and will observe the local safety regulations of their Institution. Government regulations may also apply.

We hope that the *Methods* will prove popular and welcome any feedback, in the form of new protocols or improvement of existing protocols for future editions.

Acknowledgements

We thank all who contributed methods for this section and Valerie Brame and Wendy Senior for their help in typing and editing. We would like to dedicate this volume to the The Imperial Cancer Research Fund, which supports our research.

I. M. LEIGH
F. M. WATT
London 1993

Part I

Culture techniques and grafting methods

Introduction

Scientists have been culturing keratinocytes for 100 years. In 1897 C.A. Ljunggren reported that excised pieces of human skin remained viable when cultured in ascites fluid at room temperature, and that the skin could subsequently be grafted onto wounds. During the first half of the twentieth century, as *in vitro* culture techniques were developed, experiments were carried out on explants of skin and the epidermal sheets that migrated out from them (for review see Matoltsy, 1960). Later, cultures were established from disaggregated keratinocytes, but high seeding densities were required and subculture was rarely successful (for review see Green, 1980). A major advance came in 1975, when Rheinwald and Green reported that human epidermal keratinocytes could be cultured at clonal seeding density if maintained in the presence of a feeder layer of lethally irradiated 3T3 mouse embryo cells.

The Rheinwald and Green procedure for growing human keratinocytes is still the technique of choice for a wide variety of applications (see Navsaria *et al.*, Chapter 1). It can be used to grow keratinocytes from a variety of other species, from neonatal and adult donors and from a range of stratified squamous epithelia. Large numbers of cells can be grown from small biopsies, and the technique has been used to produce grafts for burns or wounds (see Navsaria & Leigh, Chapter 11). The culture technique also provides the starting point for establishing cultures of pure melanocytes (see De Luca, Chapter 3).

A classic procedure for growing newborn mouse keratinocytes involves seeding the cells at high density without a feeder layer in medium containing a low concentration of calcium ions (see Hennings, Chapter 5). In low calcium medium keratinocytes are prevented from stratifying, because assembly of desmosomes and other calcium-dependent intercellular adhesion complexes is inhibited. Human keratinocytes can also be grown as a monolayer in low calcium medium through a simple modification of the Rheinwald and Green technique (see Watt, Chapter 2).

For some types of experiment, such as assays of growth factor production, it may be preferable to grow keratinocytes in the absence of a feeder layer in serum-free medium. We have therefore included protocols for growing human and mouse keratinocytes in defined media (see Mitra & Nickoloff, Chapter 4; Morris, Chapter 6). Media can either be obtained commercially (see Mitra & Nickoloff, Chapter 4) or made up from scratch (see Morris, Chapter 6). Both normal and psoriatic keratinocytes can be grown in defined media (see Mitra & Nickoloff, Chapter 4).

The degree of histological differentiation that is achieved when keratinocytes are grown with any of the above methods is rather limited. In order to improve morphological differentiation, keratinocytes can be grown on substrates designed to resemble the dermis. These substrates are based on gels of type I collagen and contain mesenchymal cells such as dermal fibroblasts or 3T3 cells (see Parenteau, Chapter 9; Fusenig, Chapter 10; Hansbrough, Chapter 12). The resulting keratinocyte cultures are sometimes referred to as 'skin equivalents' 'organo-typical cultures' or 'composite cultures', and they have been used for grafting both humans and animals.

This section also includes several organ culture methods. Embryonic human and mouse epidermis will stratify and differentiate in organ culture and embryonic mouse skin can go on to form hair follicles *in vitro* (see Fisher, Chapter 7). Hair follicles and sebaceous glands from adult human skin can be maintained in organ culture or can be disaggregated and used to establish cultures of follicular keratinocytes and sebocytes (see Kealey & Philpott, Chapter 8).

In summary, this Section contains a variety of methods of growing keratinocytes. The method of choice depends on the relative importance of obtaining large numbers of cells or good morphological differentiation and whether or not the presence of serum or mesenchymal cells poses a problem for your experiments.

I.1   Human epidermal keratinocytes

1   Growth of keratinocytes with a 3T3 feeder layer: basic techniques

H. A. NAVSARIA, C. SEXTON, V. BOUVARD, and I. M. LEIGH

Introduction

The technique that we shall describe was devised originally by Jim Rheinwald and Howard Green (Rheinwald & Green, 1975). Keratinocytes are seeded at clonal densities in the presence of a feeder layer of random bred Swiss 3T3 cells, which secrete both extracellular matrix proteins that encourage keratinocyte attachment, and growth factors that stimulate proliferation (Green, 1980; Rheinwald, 1980; Alitalo et al., 1982). A clone of 3T3 cells, designated J2, was selected for its ability to provide optimal support of keratinocyte growth when used as a feeder layer, although NIH 3T3 and Balb/C 3T3 lines work adequately (Rheinwald, 1989). The composition of the culture medium has undergone various improvements since 1975, and the formulation we use is essentially that of Allen-Hoffmann and Rheinwald (1984). In Rheinwald/Green, keratinocytes from neonatal epidermis can undergo up to 100 population doublings prior to senescence and keratinocytes from adults have a lifespan of 40–70 cell generations. The method can be used to grow keratinocytes from skin, oral cavity, oesophagus, exocervix, and conjunctiva, and to grow bladder urothelial cells and cells of the mammary gland.

Solutions and medium supplements

Sera
Fetal calf serum must be batch tested for its ability to support maximal colony forming efficiency and growth rate of keratinocytes. It can be purchased in large batches and then stored for 1–2 years at $-20\,°C$. Donor calf serum is used to grow the 3T3 cells and should also be batch tested.

Mitomycin C
This is prepared by dissolving a 2 mg vial in 10 ml of calcium free HEPES buffered Earle's salts (HBES).

Keratinocyte methods by Irene Leigh and Fiona Watt
© Cambridge University Press, 1994, pp. 5–12

Thermolysin
This is prepared as a 750 μg/ml stock solution in 10 mM HEPES pH 7.0; 1 mM $CaCl_2$.

Hydrocortisone
This is dissolved in 100% ethanol at a concentration of 5 mg/ml and stored at -20 °C. It is then diluted in HBES to make 40 μg/ml (100 $\times$) stock solution and stored at -20 °C.

Epidermal growth factor
This is prepared as a 1 μg/ml (100 $\times$) stock solution in DMEM and stored at -20 °C.

Cholera enterotoxin
This is dissolved at 1 mg/ml(10^{-5} M) in sterile water and stored at 4 °C. It is diluted to a final concentration of 10^{-10} M in the culture medium.

Insulin
This is dissolved at 5 mg/ml in 0.005 M HCl. This is 1000 $\times$. It is stored frozen.

Adenine
Stock solution is 1.8 mM: 24.3 mg/100 ml in 0.05M HCl. Final concentration is 1.8×10^{-4} M. It is stored frozen.

Trypsin
Trypsin (0.25%) is prepared by combining 2.5 g trypsin + 1 g glucose + 3 ml 0.5% phenol red solution in 1 litre of phosphate buffered saline (PBS) (pH 7.35). It is stored in 20 ml aliquots at -20 °C.

EDTA
0.2 g/litre of PBS is 0.02%. It is stored at room temperature.

Trypsin/EDTA
1 part trypsin + 4 part EDTA are combined for harvesting 3T3 cells and keratinocytes.

Complete media

Transport medium

DMEM containing 0.1 g/l streptomycin and 0.1 g/l penicillin. with 10% fetal calf serum and 2.5 μg/ml fungisone.

3Tβ culture medium (3T3CM)

DMEM containing 0.1 g/l streptomycin and 0.1 g/l penicillin and 5% donor calf serum.

Complete keratinocyte culture medium (KCM)

DMEM and Ham's F12 medium in a ratio of 3:1 (v/v)
5–10% fetal calf serum (heat inactivated, 56 °C for 30 min)
0.4 μg/ml hydrocortisone
10^{-10} M cholera toxin
10 ng/ml epidermal growth factor
5 μg/ml insulin

Sources

Dulbecco–Vogt Eagle's Medium (DMEM), Ham's F12, fungisone, glutamine, penicillin, streptomycin, trypsin are obtained from ICN Biomedicals (Costa Mesa, CA, USA). Mitomycin C, thermolysin, collagenase, insulin, hydrocortisone, cholera toxin are from Sigma (St. Louis, MO, USA). Epidermal growth factor (EGF) and basic fibroblast growth factor (bFGF) are from Austral Biologicals (San Ramon, CA, USA). Adenine is from United States Biochem. Corp. (Cleveland, Ohio, USA). Fetal and donor calf serum are purchased from Sera Labs. (Crawley Down, Sussex, England) Gibco BRL, (Gathersburg, MD, USA) or Imperial Laboratories (Andover, Hampshire, England.

Preparation of 3T3 feeder layer

(i) Frozen vial of 3T3 cells should be used to generate fresh stocks which are maintained in culture for 8–12 weeks. 3T3 cells should not be cultured continuously for longer than this as they tend to transform or senesce and no longer provide good feeder support for keratinocytes.

(ii) 3T3 cells are grown in 3T3CM. The cells are disaggregated by incubation at 37 °C in trypsin/EDTA. The cells are routinely split at a ratio of 1:5 two to three times a week.

(iii) To prepare feeders, confluent cultures or trypsinized cells resuspended in medium are irradiated with 6000 rads (cobalt^{60}) and replated at a third of saturation density (10^6 per 100 mm flask). The cells will attach within 2 h and keratinocytes can then be added in KCM. 3T3 cells should not be used more than 48 h after irradiation.

(iv) As an alterative to irradiation, mitomycin C can be used to inhibit division of 3T3 cells. The cells are incubated for 2 h at 37 °C in medium containing 4 μg/ml mitomycin C. The cells are washed thoroughly (three times) to ensure that no mitomycin C remains to inhibit the growth of the keratinocytes, harvested and replated as described above.

Isolation of primary cells from skin

(i)   Common sources of skin are foreskins from circumcisions, abdominal
       and breast skin from cosmetic surgery. The skin is placed in transport
       medium and moved to the laboratory, where it is processed as soon
       as possible. If necessary, the skin can be stored overnight at 4 °C;
       however, the viability declines on storage.

(ii)   Using a pair of scissors, all excess dermis and connective tissue are
       removed. For large samples, a skin graft knife or a dermatome can
       be used.

(iii) The skin pieces are washed and cut into small fragments 1–2 mm in
       size. The classical technique for isolating keratinocytes involves
       placing these fragments in a suspension flask with a side arm and a
       magnetic stir bar. 10–15 ml of a 1:4 trypsin (0.25%); collagenase
       (1%) solution is added to the flask and stirred at 37 °C. At 30–40
       min intervals, the supernatant is drawn off and new solution added.
       The tissue fragments are digested for a total of 2–3 h.

(iv)   An alternative method involves transferring the skin to a sterile petri
       dish following trypsinization and, by means of a fine pair of forceps
       or needles, physically separating the epidermis from the dermis. The
       contents are transferred into a sterile Falcon tube and resuspended
       in culture medium.

(v)   The cell suspension is filtered through a sterile gauze (or a metal tea
       strainer) into another Falcon tube. The keratinocytes pass into the
       tube, leaving solid debris trapped in the filter.

(vi)   The cell pellet is resuspended in KCM, counted in a hemocytometer
       and plated at densities of 10^5 to 3×10^5 per 1000 mm dish containing
       the irradiated 3T3s. The flask is placed at 37 °C in a 5% CO_2
       incubator. Gentamycin (50 μg/ml) and mycostatin (50 U/ml) may
       be added to discourage bacterial contamination; however, they inhibit
       keratinocyte proliferation and are not usually required.

(vii) The medium is changed two–three times weekly. It is important to
       maintain the optimal 3T3 density and so, if the feeder layer becomes
       sparse, more 3T3s should be added. Colony forming efficiency is
       0.1–2% in primary culture.

Discouraging fibroblasts

The 3T3 cells inhibit growth of human fibroblasts but, if the 3T3 density
becomes sparse, fibroblasts can become established. If fibroblast contamination is observed, the cultures are rinsed with 0.02% EDTA in PBS, then
incubated in EDTA for 30 seconds and aspirated vigorously to detach
fibroblasts and 3T3s leaving the keratinocytes attached. Fresh 3T3 cells
are then added to the cultures.

Subcultivation of keratinocytes

(i) The cultures are rinsed with EDTA to remove feeders and then incubated with trypsin/EDTA at 37 °C for about 20 min, when keratinocytes can be seen to be rounded up and detaching. The cells are transferred to a Falcon tube containing KCM to inhibit the trypsin, pelleted by low speed centrifugation (500 g for 5 min) then resuspended and replated as above.

(ii) To maintain cell stocks for experiments, the majority of cells are cryopreserved at each passage. Cells are frozen at concentrations of 10^6 or 2×10^6 cells/ml in foetal calf serum containing either 10% glycerol or 10% dimethylsulphoxide as a cryoprotectant. Freezing is performed at a rate of -1 °C per min for 30 min and then -5 °C per min to a temperature of -80 °C; at this stage the tubes are transferred to liquid nitrogen.

(iii) Ampoules of cryopreserved cells are thawed rapidly by incubation in a water bath at 37 °C. The ampoule is swabbed with 70% ethanol before opening. The cells are diluted in DMEM and 10% calf serum, then pelleted, resuspended and plated as above.

Appearance of cultures

The keratinocytes grow as tightly adherent rounded epithelioid colonies (Fig. 1.1). The growth rate slows after 8–11 days, even if the cultures

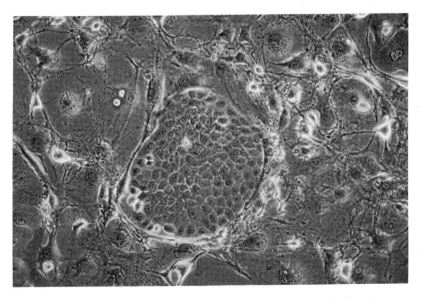

Fig. 1.1. Single colony of human epidermal keratinocytes surrounded by 3T3 feeder layer, grown using the Rheinwald and Green technique. Magnification × 120.

are not yet confluent. Thus, to maintain maximal growth, the cells should be subcultured every 8–10 days.

Cultures can be stained with 0.2% methylene blue solution following washing and 10 min fixation with 10% formalin in PBS.

Keratinocytes can be identified by reaction with antibodies to keratins, particularly highly cross reactive antibodies such as LP34 (see Rugg, Chapter 23).

Primary explant cultures

When dealing with tiny biopsies (1–2 mm) of pathological material, the explant method can be helpful to establish the primary culture: below is a typical technique.

(i)   The biopsy is removed from transport medium, transferred to a sterile petri dish and excess dermis removed. The resulting epidermis is cut into very tiny pieces with a scalpel. Fetal calf serum is added to the minced tissue (this will help to attach the epidermis to the culture dish), and the tissue is then spread out evenly over a 35 mm diameter petri dish. The sample is air-dried for approximately 4 min (do not allow to dry out completely) to attach the tissue to the plastic surface.

(ii) Keratinocyte culture medium is added to the culture together with the appropriate number of 3T3 feeders and medium changed twice weekly.

(iii) Within 1–3 weeks the keratinocytes start growing from the edges of the attached tissue. A high feeder density is maintained to discourage fibroblast outgrowth. Keratinocytes from explants can be harvested using trypsin/EDTA and replated onto a new flask to expand the population as above. The original petri dish with the explant can also be maintained to generate more keratinocytes.

Cervical keratinocytes

Cervical keratinocytes are normally obtained from hysterectomies for noncervical disease. Immediately after surgery, biopsies of cervix are placed into transport medium and stored at 4 °C for 24 h. This serves to decrease the amount of contamination in the subsequent cultures and does not substantially affect the viability of the cells.

Although the most widely used method for extracting keratinocytes for culture is through trypsin treatment, we have found that thermolysin treatment results in fewer contaminating fibroblasts from the connective tissue. This enzyme separates the epithelium from the connective tissue at the level of the lamina lucida.

Keratinocytes are isolated as follows.

(i)   Remove connective tissue with scissors and forceps.

(ii)   Wash extensively with PBS (containing 100 U/ml penicillin G; 100 μg/ml streptomycin; 2.5 μg/ml fungisone).

(iii)   Remove more connective tissue if possible.

(iv)   Wash twice with PBS (containing antibiotics), then 95% ethanol and finally once more with PBS.

(v)   Cut tissue into 0.4 cm^2 fragments and incubate in a petri dish with 750 μg/ml thermolysin for 3 h at 37 °C.

(vi)   After incubation, the very thin sheet of opaque epithelium can be easily removed from the connective tissue with forceps.

(vii)   The epithelial fragments are incubated in 1 ml of 0.25% trypsin at 37 °C for 1 h.

(viii) Add an equal volume of keratinocyte culture medium minus EGF and pipette to break up clumps of cells. Centrifuge and count the cells and plate at a density of 2 × 10^6 in a T75 culture flask on a 3T3 feeder layer prepared as above.

(ix)   Cells are incubated at 37 °C in 10% CO_2, in keratinocyte culture medium lacking EGF (Rheinwald & Green, 1977). 3 days after plating, EGF is included in the culture medium. Cervical keratinocytes will reach confluence after 10–14 days with new feeders added once per week.

Oral keratinocytes

The ease of obtaining oral, buccal and gingival tissue coupled with the rapid expansion and growth of oral cells makes oral mucosa an excellent source of keratinocytes. Contamination with yeasts is a problem intrinsic to the mouth, however. Redundant tissue from wisdom tooth extractions is placed in transport medium and stored at 4 °C from 24–46 h. Keratinocytes can be separated by serial trypsinization as above or separated from the connective tissue as follows.

(i)   Cut specimen into 1 cm^2 pieces. Wash rapidly in 70% ethanol, then in DMEM. Place pieces in a small tube containing 1 ml of 0.25% trypsin (the amount of trypsin may be scaled up to cover the quantity of tissue being prepared) and incubate for 60–90 min at 37 °C.

(ii)   Transfer to petri dish and with sterile large bore needles tease the epithelium away from the connective tissue. Care should be taken not to disturb the connective tissue as this will result in fibroblast contamination of the keratinocyte cultures.

(iii)   Pipette the trypsin solution and epithelium to create a cell suspension and plate in 25 cm^2 flask (or larger if more than 1 ml of trypsin was used) with feeder cells as for epidermal keratinocytes above.

Porcine keratinocytes

Porcine keratinocytes can be cultured for animal studies of wound healing under identical conditions to human epidermal keratinocytes (Kangesu *et al.*, 1993). Alternatively, the cells can be grown in a modified culture medium without the continuous use of a 3T3 feeder layer (Regauer & Compton, 1990*a*). Deepidermized explants of pig skin have been used as a model to study the effects of keratinocyte grafting (Regauer & Compton, 1990*b*).

2 Growth of keratinocytes with a 3T3 feeder layer: low calcium culture of human keratinocytes in the presence of serum

F. M. WATT

Human keratinocytes can be grown in medium containing a reduced concentration of calcium ions (0.1 mM compared with 1.2–1.8 mM in standard medium formulations). The cells are prevented from stratifying, because desmosomes and intercellular adherens junctions do not assemble (see, for example, Magee, Lytton & Watt, 1987; O'Keefe, Briggaman & Herman, 1987). Nevertheless, individual cells do initiate terminal differentiation within the monolayers and, on addition of calcium ions, the differentiating cells migrate off the culture dish to form a suprabasal cell layer (Watt & Green 1982; Watt 1984; Morrison, Keeble & Watt, 1988). The culture medium is the same as for standard Rheinwald–Green cultures, except that calcium is omitted from the DMEM and F12 formulations, and the fetal calf serum is depleted of calcium ions using the method of Brennan *et al.* (1975).

20 g Chelex® 100 resin (BioRad) is mixed with 500 ml distilled deionized water at room temperature. The pH is adjusted to 7.5 with hydrochloric acid. The resin is recovered by filtration and added to 50 ml serum. After mixing for 1 h at room temperature with a magnetic stirring bar, the serum is separated from the resin and simultaneously sterilized by filtration through a 0.2 μm pore size Nalgene filter. The serum can now be added to calcium-free medium.

Notes

Calcium is not the only cation to be depleted from serum by this technique.

Human keratinocyes do not survive in medium that is completely calcium free, and it may be necessary to add back calcium ions (in the form of a 1 M stock solution of calcium chloride in distilled water) to raise the concentration to 0.1 mM.

Keratinocytes and feeders can be plated out in standard medium to encourage attachment and then transferred to low calcium medium the next day.

The 3T3 feeder layer does not survive well in low calcium medium and it may therefore be necessary to use a higher plating density of keratinocytes (e.g. 10^5 per 35 mm dish) than in standard medium.

3 Growth of keratinocytes with a 3T3 feeder layer: separation of melanocytes from mixed epidermal cell suspensions

M. De LUCA

Initiation of mixed cultures of melanocytes and keratinocytes

Melanocytes are present in early cultures of keratinocytes that are derived from neonatal foreskins or from skin biopsies of healthy volunteers and cultivated on a feeder-layer of lethally irradiated 3T3-J2 cells using the Rheinwald–Green technique (Rheinwald & Green, 1975; Green *et al.*, 1979; De Luca *et al.*, 1988; see Navsaria *et al.*, Chapter 1).

The keratinocyte growth medium consists of: Dulbecco–Vogt Eagle's (DMEM) and Ham's F12 media (3:1 mixture) containing fetal bovine serum (10%), insulin (5 μg/ml), transferrin (5 μg/ml), adenine (0.18 mM), hydrocortisone (0.4 μg/ml), cholera toxin (0.1 nM), triiodothyronine (2 nM), epidermal growth factor (10 ng/ml), glutamine (4 mM), penicillin–streptomycin (50 IU/ml).

Keratinocyte cultures used for melanocyte isolation cannot be obtained from frozen keratinocyte ampoules, since melanocytes do not survive freezing and thawing procedures.

Pure melanocyte cultures

To purify normal human melanocytes, confluent primary or secondary passage Rheinwald–Green cultures are trypsinized (2–3 days after confluence) and the cell suspension is plated in the absence of feeder-layer, at a cell density of 2.5 x 10^4/cm^2 in melanocyte growth medium (MGM)-1:E-199 containing fetal bovine serum (5%), insulin (5 μg/ml), transferrin (5 μg/ml), adenine (0.18 mM), hydrocortisone (0.4 μg/ml), cholera toxin (0.1 nM), triiodothyronine (2 nM), EGF (10 ng/ml), bFGF (1 ng/ml), PMA (1 ng/ml), glutamine (4 mM), penicillin–streptomycin (50 IU/ml). G418 (100 μg/ml) is added for 2–4 days to avoid human fibroblast overgrowth.

Twenty-four h after seeding, the medium is changed and free-floating cells are removed.

After 2–4 days melanocytes, identified by their morphology and reactivity to DOPA, are further purified by differential trypsinization: melanocytes, preferentially detached at short trypsinization times (2–3 min), are collected in growth medium and plated at a cell density of approximately 5 x 10^3/cm^2. 1,3,4-dihydroxyphenylalanine (DOPA) reaction is performed as described previously (De Luca *et al.*, 1988).

Subconfluent cultures are passaged 1:3 and cultured as above. After their isolation (100% positivity to DOPA-reaction usually obtained after 2–4 passages), melanocytes are cultured either in MGM-1 or in MGM-2, which is MGM-1 without PMA. When melanocytes are isolated and cultured directly in MGM-2, the initial growth rate is slower, and the attainment of a pure melanocyte population might require longer.

Reagents

Dulbecco–Vogt Eagle's (DMEM), Ham's F12, E-199, glutamine, penicillin–streptomycin, trypsin/EDTA are obtained from ICN Biomedicals (Costa Mesa, CA, USA). Insulin, transferrin, triiodothyronine, hydrocortisone, cholera toxin, phorbol 12-myristate-13-acetate (PMA), geneticin (G418) are from Sigma (St. Louis, MO, USA). Epidermal growth factor (EGF) and basic fibroblast growth factor (bFGF) were from Austral Biologicals (San Ramon, CA, USA). Adenine is from United States Biochem. Corp. (Cleveland, Ohio, USA). Dispase II is from Boehringer Mannheim GmbH (Mannheim Germany).

4   Cultivation of human epidermal keratinocytes in serum-free growth medium

R. MITRA and B. NICKOLOFF

The following is a brief outline of the procedure used in our laboratory for the cultivation of keratinocytes from punch biopsies of normal, psoriatic, or other types of diseased skin.

Procedure

(i)   All biopsies are obtained after informed consent and with the approval of the appropriate Human Subjects Committee. Six millimetre punch biopsies of skin obtained after local anesthesia are placed aseptically into cold sterile holding medium (Earle's balanced salt solution containing penicillin 400 units/ml, streptomycin 400 μg/ml and nystatin suspension 100 units/ml) and transported to the laboratory where they are kept at 4 °C for 3–4 h to reduce the probability of contamination. Increased rates of loss of viability are observed when the biopsies are kept cold for a longer period.

(ii)   After removing as much of the dermis as possible, the 6 mm punch biopsy is bisected and placed in 5 ml of 0.3% trypsin and 0.1% EDTA (prepared in a solution of NaCl 8.0 gm, KCl 0.4 gm, glucose 1.0 gm, and phenol red 10 mg per litre, pH 7.3 referred to as GNK) for 1.5 h at 37 °C.

(iii)   After this incubation period, the specimen is transferred to a petri dish containing low calcium (0.15 mM), serum-free medium (containing EGF 10 ng/ml, insulin 5 μg/ml, hydrocortisone 0.5 μg/ml and bovine pituitary extract 0.4% v/v KGM, Clonetics Corporation, San Diego) and is held there until the separation of dermis–epidermis is initiated. No trypsin inhibitor or serum containing medium is used to neutralize the trypsin activity since their use greatly reduces subsequent passaging of keratinocytes.

(iv)   The tissue is transferred to another petri dish containing KGM and after separating the epidermis, the underneath of epidermis and upper layer of dermis, (since a considerable proportion of basal cells remained attached to the dermis), are gently scraped with a sterile scalpel blade. The suspension is immediately centrifuged at room temperature and

Keratinocyte methods by Irene Leigh and Fiona Watt
© Cambridge University Press, 1994, pp. 17–19

the pellet resuspended in 5 ml KGM with vigorous pipetting to produce a single cell suspension. The cell suspension is seeded onto two 35 mm petri dishes (usually cells obtained from one 6 mm punch biopsy are initially seeded into two 35 mm culture dishes).

(v) After 24 h, the unattached cells are removed, the plate gently washed with KGM and fresh medium is added. Generally, after 3–4 days small clusters of adherent cells are apparent, and the culture medium is changed every alternate day with subsequent appearance of progressively larger colonies of keratinocytes. Occasionally, at this stage the presence of melanocytes is apparent. Melanocytes are eventually eliminated after passaging the keratinocytes, since KGM is not optimum to support the growth of melanocytes.

Passaging of cultures

(i) When the primary cultures of keratinocytes in the dishes are about 30–40% confluent (12–15 days), they are passaged to expand the culture. Old culture medium is removed from the culture dish and the keratinocytes are dislodged by incubation with 0.03% trypsin-0.01% EDTA in GNK for 5–7 min at 37 °C using a volume of 1 ml per 35 mm dish, 2 ml per 60 mm dish, or 4 ml per 100 mm dish.

(ii) The detached cells are pipetted up and down several times and the cell suspension then diluted with KGM (1:25), pelleted by low speed centrifugation at room temperature, and resuspended in KGM for passaging of cells.

(iii) After allowing the cells to attach and spread overnight, the unattached cells ($<15\%$) and medium are aspirated and replaced with fresh KGM. The medium is replaced every alternate day and, within 7–9 days, almost 100% confluency is achieved. In general, the cell population doubles every 24–48 h until the keratinocytes reach confluency.

(iv) Upon confluency, if the cells are maintained in KGM, there is minimal stratification or obvious differentiation; basically the cells can be maintained as a confluent monolayer for an additional 5–7 days. If subsequent passaging becomes necessary, it is essential to split the culture when the confluency is about 70–75%. Normal keratinocytes can be successfully isolated and cultivated from Castroviejo keratome, 0.1 mm thick slices of the skin by trypsinization (0.3% trypsin – 0.1% EDTA) for 30–45 min at 37 °C) followed by scraping and processing as described above.

Comparison of normal and psoriatic keratinocytes

Normal cultured keratinocytes have distinctive morphology: the cells form tightly adherent, epithelioid small colonies when observed by phase contrast microscopy. Psoriatic keratinocytes in culture appear to contain two morphologically distinguishable cell types: small basaloid cells and relatively larger epithelioid cells. As the number of passages increases, the size of both normal and psoriatic keratinocytes increases and there is a decrease in proliferation rate. We are routinely capable of passaging both normal and psoriatic keratinocytes 3–4 times, which generates approximately $(20–100) \times 10^6$ keratinocytes. Such a large, uniform population of rapidly proliferating cells is ideal for a wide variety of biochemical, genetic, and molecular–biological studies of interest to investigative skin biology.

I.2   Mouse epidermal keratinocytes

5   Primary culture of keratinocytes
from newborn mouse epidermis in
medium with lowered levels of Ca^{2+}

H. HENNINGS

Introduction

Modulation of the growth and differentiation of newborn mouse keratinocytes by manipulation of extracellular Ca^{2+} levels was first reported in
1980 (Hennings et al., 1980). With low levels of Ca^{2+} (0.02–0.09 mM) in
the culture medium, keratinocytes grow as a monolayer with a high
proliferation rate, with marker proteins characteristic of basal cells. Under
these culture conditions, keratinocyte cultures can be maintained for
prolonged periods. When the calcium ion concentration $[Ca^{2+}]$ in the
medium is elevated to levels found in most culture media (1.3 mM),
proliferation ceases, followed by terminal differentiation, cornification, and
shedding of the cells from the culture dish within a few days. Alterations
of $[Ca^{2+}]$ in culture media have been used in the development of
keratinocyte cell lines (Kulesz-Martin et al., 1980; Hennings et al., 1987),
the study of desmosome formation (Hennings & Holbrook, 1983), the
induction of differentiation-associated proteins (Roop et al., 1987; Yuspa
et al., 1989), and the study of mechanisms of carcinogenesis (Kilkenny et
al., 1985; Hennings et al., 1990; Morgan et al., 1992). The procedure below
has evolved from earlier described methods (Yuspa & Harris, 1974; Yuspa,
1985) and is currently used in the Laboratory of Cellular Carcinogenesis
and Tumor Promotion, National Cancer Institute, NIH, Bethesda, MD.
The procedure described for preparation of keratinocytes from newborn
mouse skin can be modified slightly to prepare keratinocytes from adult
mouse skin (Yuspa, 1985).

Materials and Media Preparation

- Eagle's minimal essential medium with Earle's balanced salt solution,
  nonessential amino acids and L-glutamine, without calcium chloride.
  Whittaker Bioproducts, Walkersville, MD. Catalog No. 06-174D.
- Fetal bovine serum. Intergen Company, Purchase, NY. Catalog No.
  1020-75. To lower the $[Ca^{2+}]$ of the serum sufficiently to prepare
  medium with 0.05 mM Ca^{2+}, serum is treated with Chelex 100 according
  to the method of Brennan et al., 1975.

- Chelex 100, 200-400 mesh, sodium form. BioRad, Richmond, CA. Catalog No. 142-2842.
- Trypsin, 0.25%. Gibco, Grand Island, NY. Catalog No. 610-5050 AG.
- Phosphate buffered saline (PBS) without calcium and magnesium. Quality Biological, Inc., Gaithersburg, MD. Catalog No. 14-108-5.
- Penicillin–Streptomycin, Gibco, Grand Island, NY. Catalog No. 600-5140 AG.
- Nytex Gauze: 42″ Multifilament Polyester. Martin Supply Co., Baltimore, MD. Catalog No. 1781012-Z-7N-16XX.
- Preparation of media: To Eagle's minimal essential medium, add penicillin–streptomycin (0.2%) and Chelex-treated fetal bovine serum (8%). Adjust $[Ca^{2+}]$ to 0.05 mM Ca^{2+} (low Ca^{2+} medium) or 1.3 mM Ca^{2+} (high Ca^{2+} medium) by adding an appropriate volume of calcium chloride stock solution (0.28 M).

Methods

(i)   Sacrifice newborn mice (1–3 days after birth) by carbon dioxide asphyxiation and place on ice.

(ii)   Wash mice twice with Betadine, rinse with distilled water to remove Betadine, then wash twice with 70% ethanol.

(iii) Using sterile scissors and forceps, amputate tail and limbs at second joint and discard; make longitudinal incision from tail to snout; remove skin with sterile forceps.

(iv)   Stretch and flatten skin, dermis-side down, on petri dish.

(v)   Float, dermis down, on 0.25% trypsin at 4 °C for 18 h. 8 skins may be floated on 50-75 ml trypsin solution in a 150 mm petri dish. Do not allow the edges to curl under or the epidermis will not separate cleanly from dermis.

(vi)   Place skin piece, epidermis-side down, on petri dish and peel back dermis with forceps. (Fibroblasts and hair follicle cells may be prepared from the dermal sheet.)

(vii) Mince pooled epidermal sheets and stir (with a stirring bar) in a trypsinizing flask with 8% serum-containing Eagle's minimal essential medium containing 1.3 mM Ca^{2+} for 30 min at 4 °C. (Trypsin action is inhibited by the serum; stirring releases cells from the epidermal sheet mechanically.)

(viii) Filter the epidermal cell suspension through sterile Nytex gauze to remove the stratum corneum.

(ix)   Pellet cells by centrifuging at 500–800 rpm at 4 °C for 5 min.

(x)   Suspend pelleted cells in Eagle's minimal essential medium with 8% serum containing 1.3 mM Ca^{2+}.

(xi) Count cells on a hemocytometer or Coulter Counter. Typical cell yields are $(5-8) \times 10^6$ per mouse. This yield of cells is termed a mouse equivalent. To standardize plating density from one preparation to the next, more reproducible results have been attained plating the same number of mouse equivalents (e.g. 0.5 mouse equivalents per 60 mm petri dish) rather than the same absolute number of cells.

(xii) Plate cells at relatively high density: 0.25–0.5 mouse equivalents per 3 ml medium in a 60 mm dish. Using a similar plating density, adjust the volume for larger or smaller culture vessels. Cells may be plated in medium with 0.05 mM Ca^{2+}, 1.3 mM Ca^{2+}, or an intermediate $[Ca^{2+}]$ with little effect on plating efficiency. Incubator conditions should be 36–37 °C with 7% CO_2).

(xiii) After keratinocyte attachment, wash cells once or twice with PBS (without calcium and magnesium) to remove differentiated and other unattached cells, and supply medium appropriate to the experimental protocol. Rapid cell attachment after plating in medium containing Ca^{2+} at $\geqslant 0.1$ mM allows washing and the first feeding 4 h after plating. At lower Ca^{2+} levels, cells attach and spread more slowly; washing and the first feeding are carried out after overnight attachment.

(xiv) Change medium 3 times weekly. For optimum growth and for studies involving incorporation of radioactive isotopes, change the medium daily.

Acknowledgements

The following current members of the Laboratory of Cellular Carcinogenesis and Tumor Promotion have been involved in the development of the procedures described here: Andrzej Dlugosz, Adam Glick, Lenore Hart, Luowei Li, Ulrike Lichti, David Lowry, David Morgan, James Strickland, Tamar Tennenbaum, Wendy Weinberg and Stuart Yuspa.

6 Procedure for harvesting epidermal cells from the dorsal epidermis of adult mice for primary cell culture in 'high calcium' defined medium

R. J. MORRIS

I have modified the defined medium of Peehl and Ham (1980) to support the growth of adult mouse epidermal keratinocytes in the presence of 1.2 mM calcium ions (Morris *et al.*, 1991). Under these conditions, keratinocytes grow as stratified sheets containing both proliferating and differentiating cells.

Supplies

- 4–5 adult female mice (the trypsinizing procedure was devised for the thinner skin of female mice). In our hands, harvesting from fewer than 3 mice or more than 6 seems to result in a reduced yield of cells per mouse.
- 500 ml Nalgene jar for washing the mice.
- Povidone iodine surgical solution (approx. 150 ml)
- Distilled or milliQ water (for washing).
- 70% ethanol in water (for washing).
- Autoclaved harvesting instruments in a beaker of 70% ethanol: 1 pair of scissors, 2 pairs of full curve eye dressing forceps (Miltex 18-784), 1 pair of thumb dressing forceps (Miltex 6-104), 2 #4 scalpel handles, #22 sterile stainless steel blades.
- Sterile specimen cup (Fisher 14-375-147) to hold PBS and mouse skins.
- Thin 100 × 10 sterile plastic petri dishes (Baxter #D1950) for scraping.
- 50 ml sterile conical tubes (Falcon #2098).
- 5 and 10 ml Costar disposable pipettes (Costar #4051 and 4101). We need pipettes with fairly small tips to aid in the disaggregation of the cells.
- Square sterile plastic petri dishes for scraping epidermis (Baxter #D1980). The square dishes enable more forceful scraping of the epidermis than the round ones and are more stable (do not have to be chased round inside the hood).
- Sterile graduated cylinder for measuring culture medium.
- Sterile Nalgene PETG bottles (125 ml) for making cloning dilutions.

- Corning plastic ware. We use only Corning tissue culture dishes (35 mm, #2500035; 60 mm, #2501060; #25020100 for trypsinizing skins).
- Microfuge tubes for mixing cells with trypan blue.
- Pyrex funnel 50 mm top diameter with stem removed.
- 90 mm diameter Spectra-Mesh F, 70 μm pore size (Baxter #F3233-131)
- 60 ml Nalgene jar (Fisher #11815-10B) and a 1.5 inch stir bar with pivot ring (Baxter #S8311-7).

Solutions required

- Dulbecco's phosphate-buffered saline, Ca^{2+} and Mg^{2+} free, sterile. We add 100 mg/l of gentamycin (2x).
- 0.25% trypsin solution (GIBCO #610-5090) in the above PBS (approximately 50 ml per harvest).
- 'Harvesting' or 'Scraping' medium (SMEM): Ca^{2+} and Mg^{2+} free minimal essential medium for suspension culture (Whittaker #12-126B) supple-mented with 10% fetal bovine serum (Hyclone #A1111) and 2 × gentamycin (Whittaker #17-518).
- 0.4% trypan blue in 0.9% saline (GIBCO #630-5250), filter with 0.22 mm filter to remove crystals. This is necessary even if the solution is bought sterile.
- The medium for mass cultures is high-calcium, defined SPRD-111 modified from an MCDB-151 basal medium (Peehl & Ham. 1980). We make this from scratch according to their directions without stocks 4, 7, or 11 every 4 – 6 weeks. To maintain reproducible cultures, we do not store the stock solutions as they describe, but instead prepare fresh stock solutions for each lot. 500 ml PETG bottles of MCDB-151 (without stocks 4, 7 and 11) are frozen at $-20\,°C$, are stored at $-70\,°C$, and are thawed at $37\,°C$ just before mixing. The 20 additional supple-ments given in Table 1 are stored as aliquots at $-20\,°C$ and are thawed just before mixing. The complete medium is used within one week for primary epidermal keratinocytes from adult mice (primary human epidermal keratinocytes and those from neonatal mice seem not to be so fastidious as regards the freshness of the medium).
- The vitrogen–fibronectin dish-coating solution for primary epidermal keratinocytes was modified from Lechner *et al.* (1982) (see Table 2).
  Coat the dishes and then incubate them at $37\,°C$ for at least 1 h. The dishes should be used the same day they are made. They do not have to be dry.

Method of Procedure

(i) Kill groups of 4 to 5 mice (Halothane anesthesia followed by cervical dislocation), clip dorsal fur (15 to 18 cm^2; Oster Golden A5 small animal clipper with #40 blade), and place the mice in a jar (500 ml Nalgene) with enough povidone iodine surgical solution (not scrub) such as Betadine or Prepodine to cover. The stubs of fur do not have to be removed. Shake the mice well in the jar until they are completely wet. Pour off the iodine solution and rinse with distilled or milliQ water until the water is clear. Repeat this wash and rinse with more surgical solution. Rinse until the water is clear. Rinse 2 times very well in 70% ethanol. Add 70% ethanol to cover the mice and let them soak for 5 to 10 min. Pour off the ethanol. The fur will retain some of the yellow color.

(ii) Using sterile technique, remove the dorsal skins (thumb forceps and scissors) and place them in a cup of sterile Dulbecco's phosphate buffered saline (PBS, Ca^{2+} and Mg^{2+} free; #17-512Y, Whittaker M.A. Bioproducts or equivalent) with 100 mg/l (2x) gentamycin (17-518L, Whittaker).

(iii) Using sterile forceps (full curve eye dressing forceps and scalpel (#22 blade, kept in a beaker of 70% ethanol between uses), remove a skin and place it hairy-side-down in a sterile plastic petri dish). Scrape off every trace of subcutaneous tissue until the remaining skin is nearly translucent. The skins must be very thin, however, do not scrape so hard that the hairs are pulled out from the dermal surface and do not allow the skins to dry out. Put the skins back in the PBS until all have been scraped. Now flatten the skins hairy-side-up on fresh sterile petri dishes. If required to estimate the number of cells per cm^2 of epidermis, use a marker to trace on the bottom of the petri dish around the edges of the skin (include any holes or tears). Then transfer the tracing to graph paper, weigh the total tracings, and calculate the number of cm^2 from a standard curve plotted from the weights of known surface areas (see Argyris, (1976)). Otherwise, cut the skins into strips approximately 0.5 to 1 × 1.5 cm with the sterile scalpel.

(iv) Pour about 15 ml of sterile 0.25% trypsin (#610-5090PG, GIBCO) in Ca^{2+} and Mg^{2+} free PBS + 2× gentamycin into sterile plastic petri dishes. Use forceps to transfer and to float the skins hairy-side-up onto the surface of the trypsin solution. Trypsinize the skins for 2 h at 31 to 32 °C. If the mice were old or in hair regrowth, 15 min of additional trypsinizing time may be used. Insufficient trypsinization reduces the yield of viable cells; overtrypsinization results in poor growth in culture. This is a good time to coat 35 mm dishes with vitrogen–fibronectin coating, to mix the growth media, and to label

Table 6.1. Supplements to be added to one 500 ml bottle of MCDB-151[a] to make 'SPRD-111'

Supplement	Source[b]	Formulation	Amount added to 500 ml MCDB-151
1. $CaCl_2$	S # C7902	13.081 mg/ml in H_2O	5.0 ml
2. $FeSO_4 \cdot 7H_2O$	S # F8633	0.0417 mg/ml in H_2O	10.0 ml
$MgCl_2 \cdot 6H_2O$	S # M2382	12.20 mg/ml in H_2O	
3. $ZnSO_4 \cdot 7H_2O$	S # Z0251	1.176 mg/ml in H_2O	1.0 ml
4. Na_2SO_4	S # 6264	2.7 mg/ml in H_2O	650 μl
5. Trace element (100 ×)			5.0 ml
H_2SeO_3	A # 68105	see McKeehan et al., 1977	
$MnCl_2 \cdot 4H_2O$	B # 1946	($CuSo_4 \cdot 5H_2O$ was added	
$Na_2SiO_3 \cdot 9H_2O$	F # S408	in MCDB-151 stock 10)	
$(NH_4)_6Mo_7O_{24} \cdot 4H_2O$	A # 87346		
NH_4VO_3	F # A714		
$NiSO_4 \cdot 6H_2O$	S # N4882		
$SnCl_2 \cdot 2H_2O$	B # 3980		
6. MEM essential amino acids (50 ×)	W # 13-606	As supplied	10.0 ml
7. MEM vitamins (100 ×)	W # 13-607	As supplied	
8. Delipidized bovine serum albumin	CR # 40331	0.1 g/ml in H_2O	5.65 ml

9. Insulin, bovine	S # I6634 or	2.5 mg/ml in 4 mM HCl or	1.0 ml
	CR # 40305	2.5 mg/ml in H_2O	
10. Transferrin, human	S # T1147	5.0 mg/ml in H_2O	1.0 ml
11. Epidermal growth factor	CR # 40001	5.0 mg/ml in H_2O	1.0 ml
12. Glutamine	G # 320-5030	29.2 mg/ml	17.5 ml
13. Phosphoethanolamine	S # P0503	1.40 mg/ml in H_2O	2.0 ml
14. Ethanolamine	S # E9508	0.611 mg/ml in H_2O	2.0 ml
15. Penicillin–streptomycin	G # 600-5140	As supplied	5.0 ml
16. Retinyl acetate	G # 850-3000	1.0 mg/ml in absolute ethanol	57.5 μl
17. Ergocalciferol	G # 850-3301	0.1 mg/ml in absolute ethanol	500 μl
18. Linoleic acid–bovine serum albumin	S # L8384	0.1 mg/ml in H_2O	500 μl
19. Hydrocortisone	CR # 40203	1.0 mg/ml in absolute ethanol	500 μl
20. L-Cysteine HCl · H_2O	G # 810-1035	42.04 mg/ml in H_2O	500 μl

[a] MCDB-151 medium without stocks 4, 7, and 11 was made according to the direction of Peehl & Ham (1980) with reagents from GIBCO and Sigma, and freshly distilled deionized water.

[b] The abbreviations used are: A, Alfa Products (Danvers, MA); B, Baker Chemical Company (Phillipsburg, NJ); CR, Collaborative Research, Inc. (Bedford, MA); F, Fisher Scientific (Fair Lawn, NJ); G, GIBCO (Grand Island, NY); S, Sigma Chemical Company (St. Louis, MO); W, Whittaker, M. A. Bioproducts (Rockville, MD).

Table 2. *Vitrogen–fibronectin coating for Corning dishes*

MCDB-151	100 ml
Fibronectin	1.0 mg
Bovine serum albumin	10.0 ml of a 1.0 mg/ml stock solution
Vitrogen collagen	1.0 ml
2M HEPES	1.0 ml
116 mM $CaCl_2$	1.0 ml

each dish with the experiment number, the sample designation, and the date.

(v) Use sterile curved forceps to remove pieces one at a time to a sterile square petri dish (use a microfuge cap to prop up the petri dish on its lid at an angle) with about 10 ml of Ca^{2+} and Mg^{2+} free MEM Eagle (for suspension cultures, #12-126B, Whittaker) with 10% FBS (Hyclone, defined) and 2× gentamycin (called SMEM) and scrape off the epidermis with a new sterile scalpel blade (#22). Discard the dermises usually, or retain samples to confirm that the basal epidermal cells and the epithelial portion of the hair follicles are removed.

(vi) Carefully pour the epidermises into a sterile 60 ml Nalgene jar with a 1.5 in stir bar with pivot ring; rinse the petri dish with a little more medium. Bring the medium level to approximately 30 ml, cover with screw cap. Stir at 100 rpm on the Thermolyne slow stirrer for 20 min at room temperature.

(vii) Use a sterile piece of 70 μm teflon mesh (Spectra-Mesh), sterile funnel, and 50 ml capped conical tube to strain the hair and sheets of stratum corneum from the epidermal cells. Poke the hair and material on the filter around a bit with the curved forceps, then rinse with about 5 ml of SMEM, and repeat to free any trapped cells. Fill the tube with the SMEM and centrifuge at 1000 rpm, 10 min, 4 °C in the IEC-CRU-5000. If required, take a sample of the material retained on the filter for histology. The teflon mesh can be reused if it is rinsed in cold water, washed in hot water with tissue culture detergent (Lindbro 7X) and then rinsed well with distilled water.

(viii) Carefully, aspirate the supernatant, add 5 ml of cold SMEM, and resuspend the cells by triturating 25 times with a 5 ml pipette. These cells can be pipetted against the walls of the test tube to disaggregate them effectively. Also, keep everything at 4 °C. Then add 25 ml of additional SMEM and triturate 25 times with a 5 ml pipette. Make any necessary dilutions for counting the cells in a hemocytometer and for determining viability. If the dorsal skins of 4 to 5 mice were used, this is usually a 1:20 dilution with 1 ml of the original 30 mls and 19 mls of additional SMEM. Pipette forcefully 25 times. The basal epidermal keratinocytes of the mouse are 4 to 7 μm in diameter and sink rapidly, so pipette the cells immediately upon mixing to obtain an accurate sampling.

(ix)   Remove approximately 0.5 ml into a small sterile tube. Then use the 200 μl pipetman to mix gently and to measure out 200 μl of cells plus 200 μl of 0.4% trypan blue solution (GIBCO). Pipette very gently 3 times. Much more than this will decrease the viability. See notes on the use of the hemocytometer. We count all nucleated cells: all gold cells except the giant ones are scored as viable; all blue or dark grey ones are scored as blue. The large gold ones are scored as nonviable because they are known to be nonproliferative differentiating cells. Yields of viable cells should be in the range of 20 to 25 $\times$ 10^6 per dorsal skin depending on the size (usually 15 to 20 cm^2) and viabilities should average approximately 60% (most of the blue cells are differentiating). In our experience, the culture will not take well unless the yield is at least 20 $\times$ 10^6 per dorsal skin. A reason for this might be that the long-term proliferative cells are probably in the hair follicles and one has to scrape quite hard to remove them. However, this remains to be proved.

(x)   Centrifuge the cells for 10 min at 1000 rpm. For mass cultures, resuspend the cells (as described in Step (viii) in 5 ml of the medium in which they will be seeded (usually SPRD-111). Then add the appropriate amount of medium for seeding onto vitrogen–fibronectin coated dishes for experiments in biochemistry or molecular biology. We often use 4 $\times$ 10^6 viable cells seeded per 35 mm dish in 2 ml; this gives within 11 to 14 days a confluent culture with a labelling index of about 25% (one h pulse with [3H]thymidine) that is beginning to stratify. We seed each dish individually in 2 ml, pipetting the suspension twice between each dish.

(xi)   The cultures are incubated in a humidified incubator at 32 °C with 5% CO_2 in air. The medium on the mass cultures is changed the day after seeding and three times weekly thereafter.

I.3   Specialized organ cultures

7   Organ culture of fetal skin

C. FISHER

Basic organ culture methods

These methods can be used to grow skin from human and mouse embryos in organ culture. Embryonic human epidermis will stratify and differentiate under these conditions but it will not form skin appendages. Embryonic mouse skin cultured under these conditions will form hair follicles; it is usually the case that, the younger the skin is at the time of culture, the better the results will be with respect to hair morphogenesis.

Materials

- Organ culture dishes (Falcon 3037) (Becton, Dickinson, and Co., Oxnard, CA).
- Plastic sterile Petri dishes.
- Wire mesh screens to fit Falcon 3037 dishes (Becton, Dickinson, and Co., Oxnard, CA). Alternatively, screens can be easily made by cutting wire mesh into equilateral triangles (~ 24 mm/side) and bending the corners to fit the edge of the central well of the culture dish).
- Millipore Filters, HA-type, 0.45 μm, 13 mm diameter (Millipore Corp., Bedford, MA).
- Hank's balanced salt solution (GIBCO BRL, Grand Island, NY).
- Dulbecco's modified Eagle's medium (Flow Lab, McLean, VA).
- Antibiotics–100 μg/ml streptomycin, 100 μg/ml penicillin
- Fetal calf serum (GIBCO BRL, Grand Island, NY).
- Watchmaker's forceps and fine dissecting scissors.

Methods

If care is taken to sterilize instruments and area in which work is done, these procedures can be carried out at the laboratory bench. Embryonic skin from mice should be dissected from the embryo under sterile conditions. After sacrificing the pregnant animal, swab the trunk with 70% ethanol, dissect the uterus containing the embryos, and remove to Hank's (or equivalent) balanced salt solution (HBSS). Remove embryos from uterus to fresh HBSS. Dissect skin from embryos; it is helpful to use a dissecting microscope at this stage, particularly if the embryos are less than 14 d.

Keratinocyte methods by Irene Leigh and Fiona Watt
© Cambridge University Press, 1994, pp. 33–36

gestation, and/or you wish to remove a particular area such as the vibrissal pad. If the embryos are less than 14 d. gestation, the skin has not yet stratified and both the epidermis and dermis are much more fragile. After 14 d. gestation, the trunk skin can be easily removed without the aid of a dissecting microscope.

Carefully cut pieces of the dissected skin ($\sim$5 mm per side) and store until ready to use. Arrange sterile wire mesh grids, with sterile millipore filters on top, over the central well of organ culture dishes. Put a drop of culture medium (DMEM, 10% FCS, antibiotics) onto the top of the filter, and into the drop place one or two pieces of skin. Under a dissecting scope carefully arrange the skin pieces on the filter so they are flat and dermis-side-down. Gently press the edge of the pieces with the tip of your forceps to encourage the pieces to stick to the filter. When the skin is arranged, pipette culture medium into the well of the organ culture dish until it just comes in contact with the bottom of the filter. The filter and skin pieces should be wetted with the medium but not covered. Culture at 37 °C and change the medium every 3 days.

Tissue recombination

Dermal–epidermal recombinations should be allowed to adhere to each other during brief (overnight) culture periods and then grafted to a suitable site in a host animal that is either immune deficient or compatible with the tissue of origin. Since this experimental procedure causes considerable tissue trauma, the recombinants recover and grow much better as grafts than as organ cultures. Grafts should be allowed sufficient time to recover and attract blood supply, and then continue their course of development. Grafts of one or two weeks are generally sufficient for studies of mouse skin development, while studies of human skin development may require months.

Materials

- Nutrient agar, dehydrated (Difco Laboratories, Detroit, MI).
- Basal medium, Eagle's (BME), 2X (GIBCO BRL, Grand Island, NY).
- Gentamycin, 10 mg/ml (GIBCO BRL, Grand Island, NY).
- Fetal calf serum (GIBCO BRL, Grand Island, NY).
- Hank's balanced salt solution (HBSS) (GIBCO BRL, Grand Island, NY).
- Trypsin (1:250) (Difco Laboratories, Detroit, MI).

Methods

The recombinations should be allowed to re-establish adhesive connections between dermis and epidermis following overnight culture. The organ culture system discussed above is adequate for this purpose, but an agar-based semisolid medium is preferable because it also serves as a soft substrate on which to do the manipulations.

(i)   To make semisolid culture medium, dissolve and sterilize 1 gram agar in 50 ml distilled water by autoclaving. When this has cooled below 60 °C, add 50 ml 2X BME and 100 μl gentamycin. Aliquot into 5–10 ml quantities in sterile containers and store at 4 °C. Before use, melt the media by heating to 50–60 °C. Using a syringe, draw up 0.2 ml FCS, 0.8 ml of the melted media, and rapidly mix together by expelling from and re-drawing into the syringe. (This must be done fairly rapidly so the media does not begin to set.) Use this to fill the bottom of a 35 mm petri dish or use 0.5 ml in the well of an organ culture dish. Allow 30–60 min for it to set.

(ii) Dissect skin into ~ 25 mm² pieces and hold in HBSS. Dissolve trypsin (100 mg) in HBSS (10 ml) to make a 1% solution. It helps to do this an hour or two ahead of time to give the trypsin time to dissolve properly. Incubate the skin piece in the trypsin solution in the refrigerator (4 °C) for 1.5–2 h. Under a dissecting microscope, test whether the epidermis is separating from the dermis by gently teasing the two tissues apart with watchmaker's forceps or other suitable fine instruments. If they appear to be separating easily, rinse the trypsin solution off the skin pieces with HBSS containing 20% FCS. Allow the pieces to sit in this solution for 5–10 min and then begin separating the pieces of skin. The further hair development has progressed, the more care will be required to cleanly remove the hair follicles with the epidermis. As you separate the components, remove the dermal and epidermal pieces to separate storage containers (35 mm petri dishes work well) containing HBSS and FCS. Be careful to remove contaminating cellular elements from both dermal and epidermal components.

(iii) When ready to construct the recombinations, transfer the dermal pieces to the agar plate (a sterile, plastic Pasteur pipette works well) and under a dissecting scope arrange the pieces so that the side that normally faces the epidermis (papillary dermis) is up.

(iv) Transfer the epidermal pieces, which may curl quite a bit so that the apical most cell layers are facing outwards, to the agar plates and carefully move them over the surface of the agar with fine forceps. The excess fluid on the surface of the media will form a meniscus against the forceps and this can be used to float an epidermal piece over a dermal piece. Arrange the epidermis basal-side-down on the dermis, carefully remove excess fluid around the recombinants with a drawn glass Pasteur pipette, and incubate for 1–2 days at 37 °C in 5% CO_2. Graft to an appropriate host animal in a site that will allow easy recovery following an extended period of growth. The anterior chamber of the eye, or beneath the kidney capsule, are two good sites that offer a rich vascular supply and high probability for recovery of the grafts. Subcutaneous grafting is technically the easiest to perform but if the grafts are small, as generally are tissue recombinations, they may be difficult to recover.

8   Culture of human pilosebaceous units

M. PHILPOTT and T. KEALEY

Introduction

In this Section we describe methods for the isolation of human hair follicles and sebaceous glands, and their maintenance in organ culture and cell culture.

Isolation of human hair follicles

In our laboratory (Philpott, Green & Kealey, 1990) we usually isolate human hair follicles from full thickness skin taken from women undergoing facelift surgery, but we have also found that this method can also be used for the isolation of hair follicles from beard, axillary and pubic skin. It is essential, when isolating human hair follicles, that the subcutaneous fat should not be removed from the skin as this is an important factor in determining the success of our method.

The skin is cut into thin strips (approximately 10 mm × 5 mm) which are kept in EBSS:PBS$^+$ (with Ca^{2+} and Mg^{2+}) (1:1) until required for isolating hair follicles. Hair follicles are isolated by taking a strip of skin, and under a dissecting microscope cutting very carefully through the skin at the dermo-subcutaneous fat interface using a sharp scalpel blade (Swann-Morton No 12). This cuts the hair follicles just beneath the opening of the sebaceous gland resulting in the lower portion of the hair follicle, including the bulb, being left in the subcutaneous fat.

The hair follicle bulbs are then easily isolated from the subcutaneous fat by placing the fat cut side upwards under the dissecting microscope. Using watchmakers forceps (No 5), carefully grasping the outer root sheath of the cut end of the hair follicle, the hair follicle can be gently removed from the fat. Isolated hair follicles are placed in EBSS:PBS$^+$ (1:1) prior to being used for experiments. Using this method we routinely isolate several hundred hair follicles within two hours, although this is dependent upon both the age of the patient from which the skin is taken and the amount of skin used. We usually use skin within several hours of removal from the patient, and once isolated, hair follicles are left for the shortest possible time in the isolation medium before being used for experiments.

Keratinocyte methods by Irene Leigh and Fiona Watt
© Cambridge University Press, 1994, pp. 37–44

Analysis of morphology, DNA synthesis and keratin synthesis.

It can be difficult to obtain good histological sections through isolated hair follicles but we have found that placing hair follicles into agar blocks greatly facilitates subsequent handling and sectioning of the follicles. To do this we fix the follicles overnight in 3% phosphate buffered formaldehyde and then mount them in agar blocks. This is achieved by making up a 3% agar solution which is heated to 100 °C to melt the agar. A thin layer of this agar (2 to 5 mm) is then poured into a 35 mm plastic petri dish (Falcon 1007) and allowed to set. Fixed hair follicles are then aligned on the agar in small clumps (four or five follicles). Any excess liquid is blotted away with filter paper and the follicles are then coated in agar with a pasteur pipette. It is important to ensure that the agar has cooled to around 40 °C before applying to the follicles, otherwise they can be damaged. Once cool, the agar blocks containing the hair follicles are trimmed with a scalpel blade and fixed overnight in 3% phosphate buffered formaldehyde before being processed into wax for sectioning.

Patterns of DNA synthesis in isolated human hair follicles can be studied using [methyl-^{3}H] thymidine autoradiography. This is carried out by incubating four or five hair follicles in Eppendorf tubes containing 500 μl of Williams E medium supplemented but also containing 5 μCi of [methyl-^{3}H] thymidine (specific activity 3.3 μCi nmole^{-1}). Eppendorf tubes are placed, with their caps removed, inside glass scintillation vials containing 1 ml of water. The scintillation vials are then sealed with rubber Suba Seal stoppers and gassed with 5% CO_2/95% air and incubated for 6 h at 37 °C in a shaking water bath. Follicles are then washed three times with PBS$^+$ containing 10 mM cold thymidine (to remove nonspecific bound thymidine) and fixed and processed for histology as described above. Autoradiography is carried out using Ilford K5 dipping emulsion (Ilford Ltd, Mobberley, Cheshire) as previously described (Philpott *et al.*, 1990).

Patterns of keratin synthesis in isolated human hair follicles can be studied by incubating hair follicles in Williams E medium in Eppendorf tubes containing 100 μCi of 1 mM [^{35}S] methionine (specific activity 0.22 μCi nmol^{-1}) for 24 h at 37 °C. Follicles were then washed three times in PBS containing 10 mM methionine and then homogenized in ice-cold lysis buffer (1% Triton X-100, 1% sodium deoxycholate, 0.1% SDS, 50 mM NaCl, 5 mM EDTA, 1 mM phenylmethylsulphonyl fluoride (PMSF), 50 mM Tris-HCl, 30 mM sodium pyrophosphate, pH 7.4; Green *et al.*, 1986). The homogenate is then centrifuged at 12 000 g in an Eppendorf microfuge for 15 min and the supernatant discarded. The pellet is then twice extracted with a high salt buffer (600 mM KCl, 5 mM EDTA, 5 mM EGTA, 50 mM Tris-HCl, pH 7.4; Mischke & Wilde, 1987). The supernatant was discarded and the insoluble pellet analysed by sodium dodecyl sulphate (SDS)-acrylamide gel electrophoresis (Laemlii, 1970). Gels are dried under vacuum and autoradiographed using Kodak X-OMAT diagnostic film.

Maintenance of isolated human hair follicles as organ cultures in serum-free defined medium

Isolated human hair follicles are maintained in individual wells of 24 well multiwell plates (Falcon 3047) containing 500 μl of Williams E medium supplemented with 2 mM L-glutamine, 10 μg ml^{-1} insulin, 10 ng ml^{-1} hydrocortisone, 100 Units ml^{-1} penicillin, 100 μg ml^{-1} streptomycin at 37 °C in an atmosphere of 5% CO_2/95% air. Length measurements are made on isolated hair follicles using a Nikon Diaphot inverted microscope with an eye piece measuring graticule (2.5 mm). Medium is usually changed every 3 to 4 days, although we have found that hair follicles will grow normally for up to 10 days without a medium change.

Outer root sheath cell cultures

There are two basic methods that can be used for establishing outer root sheath keratinocyte cultures. One method involves establishing primary explants from either plucked or dissected human hair follicles (Weterings, Vermorken & Bloemendal, 1981; Wells, 1982) while the second involves isolating keratinocytes from the outer root sheath by trypsinization and then plating the single cell suspension onto 3T3 feeder layers (Limat & Noser, 1986).

Primary explant culture of outer root sheath keratinocytes

A number of different tissue culture media and substrates have been used for the culture of primary explants from the hair follicle outer root sheath. These range in complexity from establishing explants directly into tissue culture plastic using minimal essential medium with Hanks salts, buffered to pH 7.2 with 20 mM HEPES and supplemented with 0.6 mM L-glutamine, 15% FBS, 50 μg ml^{-1} gentamycin, 2.5 μg ml^{-1} Fungizone (Wells, 1982), to establishing explants onto bovine eye lens capsules that are mounted in specially designed culture dishes (Epicult; Sanibo, Nistelrode, Netherlands) using minimal essential medium (Eagles) (MEM) with Earles salts buffered to pH 7.2 with 25 mM Hepes supplemented with 15% FBS, 0.4 μg ml^{-1} hydrocortisone, 4 μg ml^{-1} insulin, 10^{-9} M cholera toxin, 10 ng ml^{-1} EGF and 50 μg ml^{-1} gentamycin (Weterings *et al.*, 1981).

In our laboratory, we culture outer root sheath keratinocytes either directly into tissue culture plastic or tissue culture plastic coated with collagen, which improves the rate of keratinocyte explant from the outer root sheath (Lenoir *et al.*, 1985). Hair follicles are isolated either by dissection or plucking and placed under a dissecting microscope where, with a scalpel blade, the hair follicle bulb is removed and remainder of the hair follicle cut into a number of small pieces (approximately 1 mm in length). 5 to 10 pieces of follicle are placed into 35 mm petri dishes and covered with a thin layer of culture medium consisting of keratinocyte SFM (Gibco) supplemented with 2 mM L-glutamine, 5 ng ml^{-1} EGF, 50 μg ml^{-1} pituitary extract and incubated at 37 °C in an atmosphere of 5% CO_2/95% air. After 24 h, most of the hair follicle pieces will have attached to the tissue culture plastic, and 5 ml of fresh medium can be added to the petri dishes. Under these conditions, we find that outer root sheath keratinocytes readily explant onto tissue culture plastic.

Outer root sheath culture from single cell suspensions

This is achieved by incubating hair follicles minus their bulbs in 0.1% trypsin, 0.02% EDTA in PBS for 10 min at 37 °C. A single cell suspension is obtained by vigorously pipetting the follicles in Dulbecco's modified Eagles medium (DMEM) supplemented with 10% newborn calf serum. Dissociated keratinocytes are centrifuged for 10 min at 200 g, and the cells from one hair follicle suspended in 2 ml of complete culture medium consisting of 3 parts DMEM containing sodium pyruvate and 1000 mg l^{-1} glucose, and 1 part of Ham's F12 supplemented with 10% FCS, 5 μg ml^{-1} insulin, 0.4 μg ml^{-1} hydrocortisone, 0.135 mM adenine, 2 nM tri-iodo-thyronine, 10^{-10} M cholera toxin, 10 ng ml^{-1} EGF, 50 Units ml^{-1} penicillin, 50 μg ml^{-1} streptomycin. Keratinocytes are seeded at $(2-3) \times 10^3$ cells/cm^2 on a 3T3 feeder layer prepared according to Rheinwald and Green (1975). Cultures are incubated at 37 °C in an atmosphere of 5% CO_2/95% air. Feeder layers are renewed once a week after selective removal of 3T3 fibroblasts using 0.02% EDTA.

Once primary cultures of outer root sheath keratinocytes have been established, either from primary explants or from single cell suspensions, they can be serially cultivated for 2 to 4 passages on 3T3 feeder layers (Weterings *et al.*, 1983; Limat & Noser 1986). Outer root sheath keratinocytes cultured on 3T3 feeder layers have been shown to produce the same keratins in vitro as they do in vivo (Stark *et al.*, 1987).

Germinative epithelium cell cultures

We have found that human hair follicle germinative epithelium can be cultured using the method recently published by Reynolds and Jahoda (1991) for the culture of germinative epithelium from rat vibrissae follicles. Hair follicles are isolated from human skin using the method described above. Under a dissecting microscope, the lower hair follicle bulb containing the dermal papilla is removed using a scalpel blade (Swann Morton No 15). Using fine dissecting needles (Watkins and Doncaster, Maidstone, Kent), or 25 G Microlance syringe needles, the hair follicle matrix and germinative epithelium are removed from the hair follicle bulb, leaving behind the dermal papilla and connective tissue sheath. Isolated germinative epithelial tissue is then placed in either 24 well multiwell plates or 35 mm petri dishes on feeder layers of human dermal papilla fibroblasts, prepared according to the method of Messenger (1984), and covered with culture medium consisting of MEM containing 20% FCS, 2 mM L-glutamine, $10\,\mu g\,ml^{-1}$ insulin, $5\,\mu g\,ml^{-1}$ transferrin, $0.4\,\mu g\,ml^{-1}$ hydrocortisone, $10\,ng\,ml^{-1}$ EGF, $50\,\mu g\,ml^{-1}$ bovine pituitary extract. Cultures are maintained at $37\,°C$ in an atmosphere of 5% $CO_2/95\%$ air. Under these conditions, cells explant out from the germinative epithelium forming colonies of small rounded cells. These germinative epithelial cells appear much smaller than ORS keratinocytes. Reynolds and Jahoda (1991) have reported similar observations for cultured rat vibrissae germinative epithelium.

Isolation of human sebaceous glands by shearing

To isolate human skin glands we use sagittal, frontal, midline chest skin (5 mm × 60 mm) from patients undergoing cardiac surgery (Kealey *et al.*, 1986). Before obtaining skin, ethical committee permission must be obtained, as must the consent of the patient. These are not usually difficult to obtain, as the removal of a sliver of skin from the incision neither affects wound healing nor produces a worse scar.

It is important that the incision is made with a scalpel, as diathermy may damage the glands. The sliver of skin should then be placed in a suitable buffer, such as PBS$^+$ or EBSS. If the media are bicarbonate buffered, the containers should be airtight. Bicarbonate buffered media are preferred, because they seem to promote better retention of glandular lipogenesis, possibly due to the bicarbonate dependence of lipogenesis.

In the laboratory, subcutaneous fat is trimmed from the skin and discarded. The skin is then washed 4 times in sterile EBSS to elute the bactericidal solution that surgeons apply to the skin, and then cut into small pieces (<5 mm) in length and sheared in 10 ml of PBS$^+$, using very sharp scissors, until a porridge-like consistency is obtained (bicarbonate-buffered medium cannot be used as it will be exposed to the air for some time). Samples of this suspension are then diluted in a petri dish with PBS$^+$ to facilitate microscopy, and the skin glands identified under a binocular microscope. Glands are isolated using watchmakers forceps (No 5) taking great care not to damage the glands. Isolated glands are placed in fresh EBSS until required for experiments.

It is important when using this technique to isolate skin glands to use medium at room temperature. Ice-cooled medium seems to impair glandular viability, possibly for the same reason that adipocytes are killed at $4\,°C$, namely that the intracellular fat or sebum solidifies and so ruptures the cell.

Organ maintenance of isolated human sebaceous glands

We have found that glandular lipogenesis in vitro is much greater after the overnight maintenance of the glands at 37 °C in bicarbonate-buffered medium than in freshly isolated glands and this is most likely due either to the bicarbonate dependence of lipogenesis or the temperature sensitivity of the glands. We therefore, as a routine, maintain our glands overnight at 37 °C on sterile nitrocellulose or millipore filters, pore size 0.45 μm, floating on 1ml of Williams E medium supplemented with 2 mM L-glutamine, 100 Units ml^{-1} penicillin, 100 μg ml^{-1} streptomycin and 2.5 μg ml^{-1} Fungizone, in an atmosphere of 5% CO_2/95% air. Glands to be maintained for longer periods are then transferred, after 24 h, still on their filters to Williams E medium supplemented with 2 mM L-glutamine, 10 μg ml^{-1} insulin, 10 μg ml^{-1} transferrin, 10 ng ml^{-1} hydrocortisone, 10 ng ml^{-1} sodium selenite, 3 nM triiodothyronine, trace elements mix (Gibco), 100 Units ml^{-1} penicillin, 100 μg ml^{-1} streptomycin and 2.5 μg ml^{-1} Fungizone, in an atmosphere of 5% CO_2/95% air, (Ridden, Ferguson & Kealey, 1990).

Rates and patterns of lipid synthesis in isolated sebaceous glands are measured by incubating five glands in 300 μl of bicarbonate buffered saline (120 mM NaCl, 25.6 mM NaHCO$_3$, 4.7 mM KCl, 2.5 mM MgSO$_4$, 1.2 mM KH$_2$PO$_4$ (Krebs & Henseleit, 1932) containing 2 mM [U-^{14}C] sodium acetate (specific activity 833 μCi mmol^{-1}) for 1 to 3 h. Incubations are carried out as described above for [methyl-^{3}H] thymidine autoradiography in Eppendorf tubes in a shaking water bath at 37 °C in an atmosphere of 5% CO_2/95% air. After incubation, the pilosebaceous units are washed in four changes of PBS$^+$. Lipids are extracted by homogenization using a glass–glass homogenizer into chloroform:methanol: 0.8% KCl:water (2:2:1:0.8 v/v; Bligh & Dyer, 1959). The chloroform fraction is then removed using a glass pasteur pipette and filtered through glass wool into a glass scintillation vial and then dried under nitrogen. Radioactivity is determined by liquid scintillation spectrometry. For thin layer chromatography, the dried extract is resuspended in 300 μl of chloroform:methanol:0.88% KCl (5:5:1 v/v) and spotted onto a 20 cm × 20 cm 250 μm silica gel chromatography plate and developed in four directions by four different solvents (Ridden et al., 1990). Thin layer chromatography plates are analyzed by autoradiography using Kodak X-OMAT AR film.

Sebocyte cell cultures

Primary cultures of sebocytes can be established either from isolated sebaceous glands (Xia et al., 1989; Zouboulis et al., 1991) or from trypsin treated keratotomed skin sections (Doran & Shapiro, 1990; Doran *et al.*, 1991).

Culture of sebocytes from isolated sebaceous glands

Primary explants of sebocytes can be established from sebaceous glands isolated either by shearing as described above or by dispase digestion. Dispase digestion involves cutting full thickness skin into small pieces (3×5 mm) which are then washed several times with PBS^-. These pieces of skin are then incubated in 2.4 U ml^{-1} dispase in PBS^- for 20 h at 4 °C after which the epidermis is easily removed from the dermis and, under a dissecting microscope, the sebaceous glands are isolated from the epidermis. Isolated glands are then plated on 3T3 feeder layers and covered with culture medium consisting of DMEM containing 4.5 g l^{-1} glucose and Ham's F12 medium (3:1) supplemented with 10% FCS, 100 Uml^{-1} penicillin, 100 μg ml^{-1} streptomycin, 0.5 μg ml^{-1} amphotericin B, 10 ng ml^{-1} EGF, 0.4 μg ml^{-1} hydrocortisone, 10^{-9} M cholera toxin and 3.4 mM L-glutamine. Cultures are maintained at 37 °C in an atmosphere of 95% air/5% CO_2.

Culture of sebocytes from keratotomed skin sections

An alternative method for the culture of sebocytes has been described by Doran and Shapiro (1990). In this method, 0.4 mm sections are cut using a keratotome. The top 0.4 mm section contains the epidermis and some dermis is discarded or used for keratinocyte culture. The second keratotome layer contains the sebaceous glands. This layer is then incubated in 10 mg ml^{-1} dispase in DMEM containing 100 μm l^{-1} penicillin, 100 μg ml^{-1} streptomycin and 10% FCS for 30 min at 37 °C. Following incubation the skin sections are incubated in 0.3% trypsin / 1% EDTA (w/v) in PBS for 15 min at 37 °C and then washed 3 times in PBS. The trypsinized tissue is then placed in Iscoves medium containing 2% human serum, 8% FCS, 2 mM L-glutamine, 100 U ml^{-1} penicillin, 100 μg ml^{-1} streptomycin and 4 μg ml^{-1} dexamethosone and scraped with a scalpel blade. This releases the secocytes from the skin section. Cells are then counted in a hema-cytometer and plated at a density of 2×10^4 cells/cm^2 onto 3T3 feeder layers. Cells are cultured in Iscoves medium as described above.

Isolation of the intact human pilosebaceous unit by microdissection

To isolate intact pilosebaceous units (hair follicle plus sebaceous gland) involves very careful microdissection which does, however, limit the number of pilosebaceous units that can be isolated at a single sitting. Isolation is carried out by cutting full thickness facelift skin into thin strips as described previously for the isolation of hair follicles. However, instead of cutting the skin at the dermis subcutaneous fat interface, the intact skin is placed under a dissecting microscope and using watchmakers forceps (No 5) and a fine scalpel blade (Swann-Morton size 15) pilosebaceous units are carefully dissected from the skin. Great care must be taken not to damage the hair follicle bulb or the sebaceous gland which can be difficult to distinguish against the surrounding dermal collagen. Because of this it is extremely difficult to isolate pilosebaceous units that are completely free of all collagen and adipocytes. Isolated pilosebaceous units are maintained in the same medium in individual wells of 24 well multiwell plates under the same conditions described for isolated hair follicles. Under these conditions (unpublished observations) we have found that hair fiber production is maintained for up to 7 days in culture, at similar rates to that reported by us for isolated human hair follicles. Rates and patterns of lipogenesis in freshly isolated pilosebaceous units are similar to those in freshly isolated sebaceous glands but, like the rates of lipogenesis of sebaceous glands, in organ maintenance, they decrease on culture. This model, therefore, at present offers no significant advantage over isolated hair follicles or sebaceous glands.

Isolation of the human sebaceous pilosebaceous infundibulum (sebaceous ducts).

These are the structures involved in acne. They can be isolated from nonhair-bearing facial skin taken from women undergoing facelift surgery (Guy et al., 1992). Using a keratotome, the top 0.1 mm of the skin, containing the epidermis, is removed. Then, with the keratotome set to 0.2 mm, a second layer containing the basal cells of the epidermis and the upper portion of the dermis above the sebaceous gland is removed. It is in this layer that the sebaceous ducts are located. Under the dissecting microscope, the sebaceous ducts can be easily identified as they are much larger than the ducts of the vellus follicle and lack the prominent hair of the terminal hair follicle. Moreover, the ducts also contain large quantities of sebum which appears dark on transillumination. Ducts are isolated by gentle microdissection and maintained free floating on Williams E medium containing 2 mM L-glutamine, 100 U ml^{-1} penicillin, 100 μg ml^{-1} streptomycin, 10 μg ml^{-1} insulin, 10 μg ml^{-1} transferrin, 10 ng ml^{-1} hydrocortisone, 10 ng ml^{-1} sodium selenite, 3 nM tri-iodothyronine, trace element mix (Gibco) and 10 μg ml^{-1} bovine pituitary extract. Ducts are maintained at 37 °C in an atmosphere of 95% air/5% CO_2.

Acknowledgements

We thank Robert Guy and Debbie Sanders in our laboratory for their help and advice. This research was supported by the MRC, SERC, The Wellcome Trust, Cystic Fibrosis Research Trust and Unilever.

I.4   Skin equivalents

9   Skin equivalents

N. PARENTEAU

Introduction

Organotypic culture is a form of three-dimensional tissue culture where cultured cells are used to reconstruct a tissue or organ *in vitro*, the objective being to allow the cells to exhibit as many properties of the original organ as possible. Organotypic culture allows the study of differentiation, response, and cell–cell interaction in a more tissue-like environment. One of the first organotypic methods described for skin consisted of a collagen lattice of Type I collagen populated with dermal fibroblasts over which epidermal cells were grown and was termed a 'skin equivalent' to denote its tissue-like characteristic (Bell *et al.*, 1981). Over the years, a variety of substrates or matrices have been used to promote the development of a differentiated epidermis *in vitro* (Boyce *et al.*, 1990; Régnier, Asselineau & Lenoir, 1990; Triglia, 1991). This section will focus on methods used to develop an *in vitro* epidermis using a dermal fibroblast-populated collagen lattice. While the techniques described are specific to this model, many of the technical notes apply to organotypic culture in general.

Method

Preparing cell stocks

It is advisable to standardize as many of your starting materials as possible since you will be doing rather complex tissue culture where events can be more difficult to interpret and problems more difficult to pinpoint. Your most important and influential starting materials are your cells. You will be using both dermal fibroblasts and keratinocytes. Both these cell types are readily obtained from normally discarded infant foreskin tissue. A number of methods exist for the primary culture and serial cultivation of both these cell types. The method used by our laboratory (Johnson *et al.*, 1992) is detailed below.

Keratinocyte methods by Irene Leigh and Fiona Watt   45
© Cambridge University Press, 1994, pp. 45–54

Establishing cultures of dermal fibroblasts and keratinocytes from infant foreskin

Materials

- Antimicrobial wash solution: phosphate buffered saline containing 100 μg/ml gentamicin; 250 μg/ml amphotericin B.
- Enzyme solution: phosphate buffered saline containing 2 mg/ml trypsin, 5 mg/ml collagenase, 50 μg/ml gentamicin sulphate and 1.25 meq/ml amphotericin B.
- 20 ml sterile filtered 95% ethanol in a 100 mm dish.
- Two sterile forceps.
- Sterile scalpels or scissors.
- 100 mm tissue culture dishes, collagen coated (see (vi)(b)).
- 100 mm bacteriological or tissue culture dishes.
- T-75 tissue culture flasks.
- Sterile capped polystyrene round-bottomed tube 16 mm/125 mm.
- Sterile magnetic micro-stirbar (flea).
- Magnetic stirplate in a 37 °C chamber.
- Dulbecco's modified Eagle's medium (DMEM) containing 10% newborn calf serum (NBCS) (prescreened for cell growth).
- 500 μg/ml trypsin-200 μg ml versene.
- Minimally supplemented basal medium (MSBM, Johnson *et al.*, 1992).

Which consists of a base of calcium-free DMEM (4.5 g/l glucose) and Ham's F12 mixed 3:1 supplemented with:

- 10 ng/ml epidermal growth factor
- 5 μg/ml bovine insulin
- 0.4 μg/ml hydrocortisone
- 20 pM triiodothyronine
- 5 μg/ml transferrin
- 10^{-4} M ethanolamine
- 10^{-4} M phosphorylethanolamine
- 5.3×10^{-8} M selenious acid
- 0.18 mM adenine
- 10^{-3} M strontium chloride
- 7.25 mM L-glutamine
- 50 μg/ml gentamicin (recommended for primary cultures).

Preparation of tissue

(i)   Wash foreskin in multiple changes of the antimicrobial wash solution
(ii) Wash foreskin for 1 minute (no longer) in the ethanol with constant agitation using sterile forceps.

(iii) Immediately transfer foreskin to fresh antimicrobial wash to remove the ethanol.

(iv) Trim the foreskin of subcutaneous tissue, wash.

(v) Transfer to fresh antimicrobial wash in a 100 mm dish and mince into 1–3 mm^2 pieces using a sterile scalpel or scissors and a fresh pair of sterile forceps.

To obtain fibroblasts

(vi)(a) Place the microstir bar into the round bottom tube and add 5 ml of the enzyme solution.

(vii)(a) Transfer approximately 1/4 of the minced tissue to the tube. Incubate the capped tube, gently stirring at 37 °C. Shake the tube vigorously after 20 min to prevent clumping of the tissue.

(viii)(a) After 30 min allow tissue to settle briefly and remove as much of the enzyme solution as possible and discard. Add 4 ml of fresh solution and continue incubation as above, shaking tube periodically.

(ix)(a) After an additional 30 min, allow tissue to settle and transfer the solution to a sterile 15 ml conical tube.

(x)(a) Add 5 ml of DMEM-10% NBCS to the conical tube containing the supernatant fraction, centrifuge for 5 min at 600 × g, aspirate the supernatant, resuspend pellet in 2 ml DMEM–10% NBCS and place on ice.

(xi)(a) Repeat steps (viii) through (x) until all the tissue is digested and only squamous sheets of stratum corneum remain. This usually takes approximately 180 min.

(xii)(a) Pool the cell fractions, count cells and determine viability. Seed the cells at 1.0×10^6 cells/T-75 flask in 10 ml of DMEM–10% NBCS. Incubate at 37 °C, 10% CO_2. Cells are passaged at confluence at 1×10^6 cells/T-150 flask in DMEM–10% NBCS.

To obtain keratinocytes

(vi)(b) Collagen coat 8, 100 mm tissue culture dishes by incubating with 5 ml of a 40 µg/ml solution of bovine tendon or rat-tail tendon acid-extracted collagen for 30 min at room temperature. Aspirate the collagen solution and wash with sterile Milli-Q or distilled water. *Do not* remove the water until use.

(vii)(b) Aspirate the water from the collagen-coated dish and, using sterile forceps, distribute 8–10 pieces of minced tissue onto each tissue culture dish placing the explants approximately 1 cm apart dermis side down. Place dishes in the incubator for 30 min to allow the tissue to stick to the culture dish.

(viii)(b)   Gently add 10 ml of MSBM to each dish of explants and incubate
at 37 °C, 10% CO_2, feeding every 2–3 days until the explant
outgrowths have begun to merge. Cells are then routinely passaged
onto collagen-coated dishes at $(1–2) \times 10^5$ cells/60 mm dish
or $(3–7.5) \times 10^5$ cells/100 mm dish in MSBM.

To generate frozen cell stocks

It is advisable to establish frozen vials of primary cells (counted as passage
1) and use a portion of the cells to generate frozen working cell stocks at
passage 3 for neonatal foreskin keratinocytes and passage 5–6 for fibroblasts
for routine use in organotypic cultures. Fibroblasts may be frozen using
routine methods however keratinocytes are frozen in DMEM containing
10% DMSO and 0.5% casein (Johnson *et al.*, 1992) to avoid exposure to
serum).

Constructing an organotypic culture

The steps are diagrammatically outlined in Fig. 9.1.

Formation of the dermal collagen lattice

Materials
- A confluent culture of dermal fibroblasts, fed the day before.
- Sterile phosphate buffered saline.
- 6-well deep tissue culture tray containing special resin tissue culture
  inserts with a 3 µm porous polycarbonate membrane*
- 10 × minimum essential medium with Earle's salts
- Gentamicin sulphate (50 mg/ml)
- NBCS
- L-glutamine (200 mM)
- Sodium bicarbonate (71.2 mg/ml).
- 500 µg/ml trypsin-200 µg/ml versene.
- DMEM-10% NBCS.
- Sterile bovine tendon or rat tail acid-extracted collagen (1.0–1.3 mg/ml)
  in 0.05% acetic acid.

Method

The following volumes are sufficient to make 6 lattices using $2\,\text{cm}^2$
culture inserts.

---

* The specialized materials used in our laboratory for the construction of organotypic cultures
are available through Organogenesis Inc. The use of a culture insert is recommended as it
allows for controlled lattice contraction, prevents epiboly of the epithelium and facilitates
growth and maintenance at the air–liquid interface.

Constructing a 'skin equivalent'

Casting the dermal layer

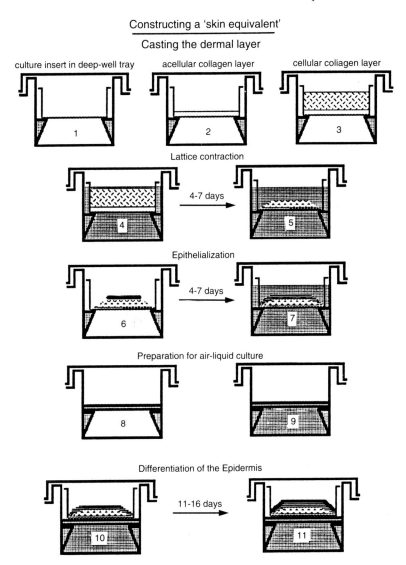

Fig. 9.1. Diagrammatic outline of skin equivalent protocol. 1. Culture insert in special deep well tray. 2. A 1 ml layer of acellular collagen is cast over the polycarbonate membrane of the insert. 3. A 3 ml cellular collagen layer is cast onto the acellular gel. 4, 5. The collagen gel contracts away from the sides of the insert to form a contracted collagen lattice having a central raised area. 6. A suspension of epidermal keratinocytes is seeded onto the central area. 7. The epithelial sheet is allowed to develop submerged. 8. 2 cotton pads are added to a deep well tray. 9. medium is added just to the level of the pads. 10. The developing culture is placed on top of the pads and cultured for the remainder at the air–liquid interface.

(i)   Make the collagen premix solution by combining in the order given:

- $10 \times$ MEM   2.2 ml
- L-glutamine   0.2 ml
- Gentamicin sulphate   0.025 ml
- Sodium bicarbonate   0.7 ml
- NBCS   2.5 ml

Place on ice.   5.6

(ii)   Measure 18.5 ml of the collagen solution (larger amounts may be aliquoted by weight for accuracy) and place in a 50 ml conical tube on ice.

(iii)   Prepare a suspension of the dermal fibroblasts at a concentration of 2.5×10^5 cells/ml in DMEM–10% NBCS.

(iv)   Add the entire volume of the collagen premix solution to the tube containing 18.5 ml of collagen. Mix well by swirling and immediately pipette 1 ml of the neutralized collagen solution into each culture insert. Place the tube on ice. Allow the acellular layer to gel at room temperature with no agitation.

(v)   Add 2.0 ml of the fibroblast suspension to the remainder of the neutralized collagen solution, swirl to mix and immediately pipette 3 ml of the cellular collagen solution into each culture insert. Allow it to remain in the hood until gelled with little or no movement.

(vi)   Add 13 ml of DMEM–10% NBCS to each well and incubate at 37 °C, 10% CO_2 until the lattices have contracted away from the sides of the culture insert to form a central raised area approximately 12–16 mm in diameter. (Note: if you are using standard polystyrene culture inserts, the collagen gel may stick to the sides of insert and must be released from the sides of the well by reaming with a sterile instrument to allow the contraction to proceed.) Contraction should take between 4 and 7 days. You do not need to change the medium during this time.

Epidermalization of the collagen lattice

Materials

- Epidermal cell cultures near confluence
- Minimally supplemented basal medium† (MSBM) (*without* EGF and strontium; *with* 1.88 mM calcium chloride, 2×10^{-9} M progesterone* and 0.3% NBCS).
- NBCS
- 500 μg/ml trypsin-200 μg/ml versene
- 2.5 mg/ml soybean trypsin inhibitor (SBTI)
- Contracted dermal lattices.

Method

(i) Trypsinize epidermal keratinocytes quenching trypsin activity with an equal volume of SBTI. Centrifuge and resuspend at $(1.8-3.6) \times 10^{-6}$ cells/ml in MSBM.

(ii) Aspirate the medium from the lattices.

(iii) Pipette 50 μl of keratinocyte cell suspension onto the central raised area of each lattice.

(iv) Incubate for 2 h at 30 °C, 10% CO_2.

(v) Gently add 13 ml of the modified MSBM into each well.

(vi) Feed after 2 days with modified MSBM.

(vii) At day 4, stain one well with 1:10 000 Nile blue sulphate in phosphate buffered saline (PBS) or DMEM to determine epithelial coverage. Stain for 30 min at 37 °C and wash with PBS. The epithelium should appear as a dark blue sheet covering the lighter staining lattice. Cultures may be cultured submerged additional days if coverage of the central raised area is less than 70%.

Differentiation of the epidermal layer at the air-liquid interface

Further differentiation of the epidermal layer requires culture at the air–liquid interface.

Materials

- Deep 6-well culture tray
- 12, 2.86 cm diam. sterile cotton pads
- Sterile forceps
- 6 skin equivalents in culture inserts with a confluent or near confluent epithelial layer covering the central raised area.
- Cornification medium: DMEM-F12† mixed 1:1 supplemented as for MSBM (above), *without* epidermal growth factor, strontium and progesterone; *with* 2% NBCS.

Method

(i) Using sterile forceps, place two cotton pads on the supports in each well of the deep tray (refer to Fig. 1).

(ii) Add 9 ml of cornification medium to each well.

(iii) Using sterile forceps, transfer the cultures to the new tray being careful to assure good contact between the pad and the culture insert. Check for air bubbles and repeat placement if necessary. The level of the medium should be slightly below the level of the culture insert. The cotton pads allow you to lower the level of the medium to ensure a dry air–liquid interface while maintaining good contact of the culture

---

* May be omitted.

† Refer to technical notes.

to the medium*. The deep tray allows for maintenance of an adequate volume of medium at this stage.

(iv) Incubate cultures at 36 °C, 10% CO_2. Change the medium after 3–4 days using 7 ml† of fresh medium.

(v) After 7 days (second media change following culture at the air–liquid interface) reduce the serum content of the medium to 1% (maintenance medium) and maintain the cultures in this medium from then on changing the medium every 3–4 days.

Technical notes

If you use a different method of keratinocyte culture, you should adjust for relative colony forming efficiency (MSBM keratinocytes can typically be 30% or higher) and modify the protocol accordingly, particularly during the epithelialization phase. For example, if you wish to use keratinocytes cultured using a 3T3 feeder layer as your starting epithelial cell population, it is advisable to increase the cell keratinocyte number approximately 2-fold at epithelialization.

The medium formulations given above were developed for use with MSBM-cultured keratinocytes. The addition of EGF and stimulatory agents such as cholera toxin should not be necessary since you are working with a high density culture. However, if colony-forming efficiency is low due to culture conditions, passage level, age of donor, etc, these agents may aid in establishing the epidermal layer. Once established, it may be possible to remove these agents. The exact media formulation used at each step will often change the timing of epidermal development rather than the final outcome (Fig. 9.2).

High glucose DMEM (4.5 g/l) is used in MSBM for keratinocyte growth and may be used to make modified MSBM for skin equivalent cultures. Alternatively, DMEM *without* glucose may be used to obtain a low glucose medium for skin equivalent culture (from epidermalization). The low glucose formulation can inhibit migration of the confluent epidermal layer on the collagen lattice, improving the stability of the cultures at the air–liquid interface. It can also enhance the formulation of the granular layer when used in the deep 6-well tray configuration. However, effects of glucose concentration will vary with the general health (metabolic rate) of the cultures and the proportion of medium to tissue. For example, the high glucose formulation works equally well when using a larger 100 mm

---

* If the medium level is higher than the base of the culture insert, you will achieve a moist interface which can inhibit full epidermal differentiation. If the medium level is too low, a meniscus can form trapping air under the culture, limiting access of nutrients. The wicking nature of the cotton pad helps ensure contact between the culture and the medium. The thickness of the pads allows for some margin of error in effective medium volume.

† The pads will retain approximately 2 ml of medium.

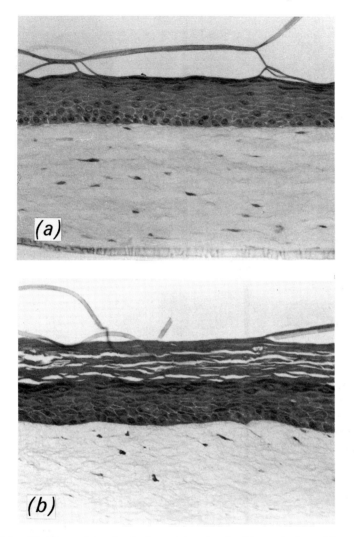

Fig. 9.2. Histology of the developing epidermis of a skin equivalent. (*a*) Cultured for 7 days at the air–liquid interface. (*b*) Cultured for 14 days at the air–liquid interface.

diameter configuration (Wilkins *et al.*, 1994) Glucose concentration should be considered if the proportion of tissue to medium differs from that described and/or problems of poor development and excessive epidermal migration occur.

If the epithelial cells have difficulty growing out on the lattice surface, they may be cultured for the first two days in MSBM containing only 0.08 mM calcium (standard) to allow migration. For example, this is beneficial

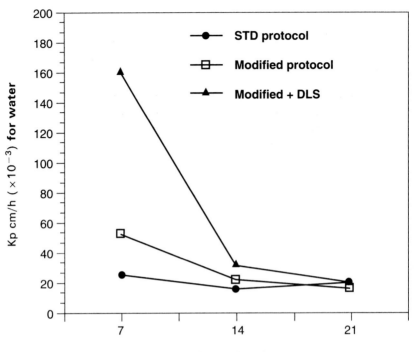

Fig. 9.3. Time course of barrier function development. The passage of tritiated water was determined at different time points during the course of epidermal development. STD protocol=standard protocol. Modified protocol included the addition of lipid supplements: linoleic acid/BSA complex (2 μg/ml), serine (1 mM), choline chloride (0.64 mM) and ethanolamine/phosphoethanolamine (0.5 mM) and modified protocol using delipidized serum (DLS). Note that the timing of barrier development is changed yet the ultimate result is the same. Histological appearance was also similar at 14 days. (Reprinted from Bilbo *et al.* (1993) *J. Toxicol-Cutaneous & Ocular Toxicol.* **12**(2) 183–96 by courtesy of Marcel Dekker, Inc.)

when epidermalizing large areas such as in the case of a sheet for grafting where even coverage over a wide area (9.6 cm & 19.2 cm) is desired (Parenteau *et al.*, 1991).

The culture surface should appear dry after approximately the first week at the air–liquid interface. If it appears moist, either the medium level is too high and/or the cultures have not continued to develop properly. Histological analysis of the cultures after approximately 7 days and 14 days at the air–liquid interface is recommended (Fig. 9.3). Depending on your starting cell population, and exact medium composition, the epidermis of your culture may look more or less developed than that shown in Fig. 3. However, when using neonatal foreskin keratinocytes, you should observe an epidermal layer exhibiting all themorphological layers of human skin after approximately 11–16 days.

Summary

Dependable construction of organotypic cultures requires consistency of materials and culture procedures. Once these are established however, skin equivalents can readily provide the investigator with a valuable intermediate step between a culture dish and *in situ*.

10   Culture of keratinocytes on collagen gels and use of transplantation chambers for grafting onto mouse skin

N. FUSENIG

Introduction

These protocols can be used to grow keratinocytes (human or mouse, normal or neoplastic) on collagen gels in order to study keratinocyte interactions with mesenchymal cells. The cultures can be grafted onto mice using a special transplantation chamber.

Preparation of collagen gels

Isolation of collagen type 1

Generally, collagen (type 1) is isolated from different tissues in the laboratory. The procedure is easy and cheap, although reproducible quality and concentration of solutions are not always obtained. By standardizing the procedure, almost pure collagen type 1 can be obtained from rat tail tendons, as proven by gel electrophoresis. Reproducible concentrations of the collagen solution can be easily obtained by using lyophilized material and adjusting the desired concentration on a dry weight basis. Alternatively, commercially available collagen preparations are equally acceptable, provided the collagen concentration is high enough for the desired gel consistency. Otherwise, the collagen solution can be concentrated by lyophilization and readjustment to the desired concentration.

    The following procedure, a modification of several published versions, is used in our laboratory.

(i)   Rat (or mouse) tail tendons are isolated by stepwise breaking the tail vertebra and removing the tissue from the tendons, starting at the base of the tail; for sterilization the (frozen-stored) tails are pretreated with 70% ethanol for 1 h.

(ii)   After elimination of blood vessels by fine forceps (the predominant source of impurities in the collagen preparations), the tendon bundles are washed in distilled water, cut with scissors into small (5 to 10 mm) pieces and dried on filter paper.

(iii)   The material is weighed and a 100 fold (ml/g) ice-cold 0.01% acetic acid solution is added for solubilization of the collagen; the vessel is placed for 24 to 48 h at 4 °C on a magnetic stirrer with slow rotation (with increasing viscosity, speed has to be adjusted).

(iv) The turbid solution is cleared by centrifugation (30 min at 30 000 g), lyophilized, and stored in tightly sealed vessels. All steps in this procedure are performed under aseptic conditions, so that further sterilization procedures are not necessary. This preparation consists of nearly pure type 1 collagen (provided the blood vessels have been extensively eliminated) as monitored by gel electrophoresis.

Preparation of collagen gels

(i)   The collagen is resolubilized at the desired concentration (usually between 2 mg/ml and 4 mg/ml) in 0.01% acetic acid and kept at 4 °C. Collagen solutions above 6 mg/ml are very viscous and difficult to handle.

(ii) The collagen solution can be gelled by exposing it at room temperature to 0.25% ammonia vapor (in a desiccator) for 20 min, followed by thorough washing in culture medium. Alternatively, and most commonly, gelation is achieved by equilibrating the pH of the solution to 7.4, adding Ca^{2+}, and raising the temperature to 37 °C as follows. Ice-cold solutions are mixed with the following quantities: 8.0 ml of collagen solution (e.g. 4 mg/ml) in acetic acid + 1.8 ml of culture medium + 0.2 ml 1 M NaOH. This gives 10.0 ml of collagen medium mixture with a final concentration of collagen of approximately 3.2 mg/ml.

(iii) Keep solution on ice and disperse, with cooled pipettes, 0.5 to 1.0 ml into 20 mm diameter wells of a multiwell plate (placed on ice).

(iv) For gelation the plate is placed for 1 h at 37 °C in a humidified incubator.

(v)   The gels can now be easily detached by fine forceps and placed on grids or any other support. Before use as culture substrate, gels should be equilibrated for salts and media components by rinsing in complete culture medium. Filters or any other surfaces can be coated with the collagen gel with the solution from step (ii).

For embedding mesenchymal cells in the gel matrix, an appropriate number of fibroblasts or other cells (5×10^4 to 1×10^5 cells/gel) are added, together with the culture medium (adjusting salt solution to isosmolality) during mixing of the gelation solution.

Mounting of collagen gels

Gels can be placed on metal grids or filters to provide air exposure of the epithelium that is grown on the gel surface. In order to prevent gel contraction by the fibroblasts and to achieve a well-delineated culture surface for the plated keratinocytes, more reproducible results are obtained by mounting the gels between the teflon rings of the Combi Ring Dish (CRD) of Noser and Limat (1987) (Distributed by Renner AG, 6701 Dannstadt, Germany).

(i) Freshly prepared collagen gels (of 20 mm diameter) with or without fibroblasts are detached from wells, placed on a piece of polypropylene film with a central perforation and carefully positioned above the outer ring of the CRD.

(ii) With the inner ring, the collagen gel and the film are pressed into the outer ring so that the gel is tightly mounted between both rings, forming the bottom layer of the inner ring.

The collagen gel can be further stabilized by cross-linking with glutaraldehyde (1% in PBS pH 7.4) for 4 h at room temperature. The gel has to be washed thoroughly in PBS and complete culture medium to eliminate all fixative in order to achieve similar cell attachment and growth as on native collagen gels.

(iii) Epithelial cells are plated (in 100 μl medium) into the inner ring of the CRD, and after attachment, the chamber is lifted to the air–medium interface by placing on Stanzen culture dishes or metal grids and adjusting the medium level just touching the lower surface of the collagen gel.

At desired time points, gels are either fixed in formaldehyde (for histology) or frozen in liquid nitrogen-cooled isopentane (for immuno-histochemistry). In order to maintain the close apposition of the different components, and to prevent damage to the epithelium during the procedure, the cultures are cut out of the chambers and embedded in an agar solution (2%) and then processed *en bloc*.

Preparing of epidermal mesenchymal organotypic cocultures (Dermal equivalents)

(i) Solutions of collagen (in 0.01% acetic acid) from various sources are adjusted to the desired concentration (2–4 mg/ml) by dilution or concentration (e.g. by lyophilization).

(ii) The collagen solution is adjusted to pH 7.4 (see protocol for preparation of collagen gels) and mixed with complete culture medium containing the desired mesenchymal cell concentration on ice (to prevent premature gelation) and pipetted into petri dishes or multiwell plates of appropriate sizes. Bacterial dishes should be used, or repopulated gels should be detached from the surface after gelation to prevent growth of fibroblasts on the petri dish and to facilitate contraction of the gel.

(iii) Contracted (after 2 to 5 days) or freshly prepared fibroblast-repopulated collagen gels are placed on metal grids (which will retard or prevent contraction) or gels are mounted in CRDs as described above.

(iv) Epithelial cells can now be plated on the upper surface of the lattices and, after cell spreading (1–2 days), the culture is lifted to the air–liquid interface by being placed on metal grids or Stanzen dishes, or is cultured submerged in medium for different time periods (usually one week) before lifting, or is left submerged for the time of the experiment. In air-exposed cultures, the medium volume is reduced in the culture dish to adjust the meniscus to the desired level (usually just above the lower surface of the collagen gel).

With squamous epithelia such as epidermal cells, medium can be completely withdrawn from the epithelial surface if cultures are in appropriate closed chambers (such as CRD) and the cells exposed to air. The use of CRD or similar devices, however, provides the modification and analysis of two culture compartments separately. Moreover, it is possible to investigate in pharmacological studies the different effects on cell physiology of agents applied either 'topically' (to the upper surface of the epithelial cultures) or 'systemically' (added to the culture medium and getting in touch first with the basal-layer epithelial cells, following diffusion through the collagen gel).

Antipodal (opposite) cocultures without contact between epithelial and mesenchymal cells

(i)   Native collagen gels are mounted within CRDs or prepared in multiwell dishes.

(ii) When mounted in CRDs, chambers are turned upside-down and placed in petri dishes, and culture medium is filled up to the exposed surface of the collagen gel (the lower surface of the upright chamber). By this means, leakage of plated cells by capillary forces through the cleft between the teflon rings is prevented. Fibroblasts are plated on the exposed lower surface ($0.5\ cm^2$) in 10 μl complete culture medium. Chambers are transferred to the incubator for attachment and spreading of the fibroblasts which is usually achieved within 1 to 2 h.

(iii) The chambers are then turned upright and placed in Stanzen dishes (or on grids) with complete culture medium added to the upper level of the collagen gel. Epithelial cells can now be plated on the upper (antipodal) surface of the gel (exposed inside the chamber) and after attachment left submerged in medium or exposed to air.

(iv) When gels are used on metal grids and not mounted in CRDs, fibroblasts are plated on freshly prepared collagen gels in multiwell dishes and after attachment and spreading, gels are carefully detached, turned upside down (the fibroblasts facing the petri dish) and keratinocytes plated on the opposite (antipodal) side of the gel. After their attachment and spreading (usually overnight) the gels are carefully placed on metal grids (or alternatively on filters placed on metal grids) with the keratinocytes on the upper and the fibroblasts on the lower surface of the gel. These antipodal organotypic cocultures can now be further incubated in medium or air exposed with the medium meniscus adjusted to the lower level of the collagen gel. Fibroblasts will migrate into the collagen gel but, depending on thickness, will not reach the epithelium formed by the keratinocytes growing at the upper surface within two weeks.

Surface transplantation method

Organotypic cultures mounted in CRDs either growing on cell-free collagen gels or gels repopulated with mesenchymal cells are prepared as described on p. 57. For transplantation of this device the (CRD) chambers are covered by the silicone transplantation chamber and transplanted *in toto* onto the subcutaneous tissue of syngeneic (mouse cells) or nude mice. This makes possible the analysis of cell behaviour under identical geometrical conditions both *in vitro* and *in vivo*.

(i) When covering the CRDs with the silicone transplantation chamber (provided by Renner KG, 6701 Dannstadt, Germany), care should be taken not to damage the collagen matrix by mechanical pressure or by creating overpressure inside the chamber (because of the tight fitting of the silicone chamber to the CRD). This can be prevented by sticking a cannula into the dome of the transplantation chamber and lifting the chamber with this cannula onto the CRD, pushing it gently down by mild pressure with forceps.

(ii) This whole assembly can now be transplanted on the subcutaneous tissue (muscle fascia) of nude mice after a sagittal incision in the back skin, insertion of the chamber and closing the skin above the rim of the transplantation chamber by wound clips.

Alternatively, a granulation tissue can be prepared prior to transplantation in order to improve the wound bed's qualities. This is achieved by implanting a rough-surfaced glass disk under the skin of mice 3 to 4 weeks before transplantation. The glass disk is introduced under the skin via an insertion close to the tail base and left in place. Within 3 to 4 week (depending on mouse strain, nude mice usually produce less granulation tissue) a foreign body granulation tissue is formed around the glass disk being well-vascularized. Following elimination of the glass disk (after incision of the skin above the disk), the transplantation chamber is placed inside the formed granulation tissue 'pocket' and the skin closed around the chamber.

(iii) At desired time intervals, the chambers with the underlying mesenchymal tissue are dissected *en bloc*, the inner part of the chambers punched out and either fixed for histology or frozen for cryostat sectioning.

I.5 Keratinocytes for grafting

11 Keratinocyte sheets for grafting

H. A. NAVSARIA and I. M. LEIGH

The expansion potential of keratinocytes cultured by the Rheinwald and Green technique has been exploited in order to graft patients with very extensive and severe life-threatening burns (O'Connor *et al.*, 1981; for review see Leigh *et al.*, 1991) and remains the gold standard for cultured graft preparation. When confluent cultures of keratinocytes are treated with the neutral protease, Dispase, the keratinocytes detach as an intact sheet from the underlying tissue culture plastic; intercellular desmosomal connections are not disrupted. The detached sheet is thin and difficult to handle; therefore, a backing dressing is applied before use to ensure correct orientation, the basal surface of the basal cells being placed in contact with the wound bed.

Patient samples

A 1 cm^2 site is excised as soon as possible and sent to the laboratory in transport medium. In very severely burned individuals, sites such as the groin, scalp and sole may have to be used. Keratinocyte expansion will take approximately 3 weeks for primary skin grafts to be prepared (up to 1 m^2) but thereafter grafting can be provided by subcultures of the primary sample. The number of cultures required is estimated. Planning of the surgical schedule and close liaison with the surgeons is required. Prior to receiving grafts of cultured keratinocytes, the patient is then managed with best practice surgical techniques, which involve covering as much surface area as possible with meshed autografts and cadaver or relative allograft.

Note on laboratory handling

In the sick patient, it may not be possible to assess risk factors for infection. Human tissue and blood products are known to harbor infectious agents such as HIV and hepatitis B virus and therefore the handling of potentially infected tissue represents a health risk to the laboratory worker. Such risks can be minimized by safe working practice, vaccination and viral screening assays. Precautions include wearing appropriate laboratory coats and disposable gloves, working in a class II safety cabinet, minimizing the

Keratinocyte methods by Irene Leigh and Fiona Watt
© Cambridge University Press, 1994, pp. 63–65

generation of aerosols, restricting the use of sharp instruments and proper disinfection of work surfaces, pipettes, flasks, etc. While no cases have yet been documented for infection of keratinoctyes by HIV, the possibility of latent infection cannot at this stage be excluded. This situation is particularly relevant when keratinocytes are to be used for allografting and may endanger the graft recipient if allogeneic infected tissue is used. Screening of the culture supernatant for the presence of HIV reverse transcriptase coupled with the use of PCR to screen isolated cellular DNA for the presence of integrated provirus effectively reduces the risk of infection to within the limits of sensitivity of the assay.

Preparation of the keratinocytes

(i)   Cultivation of keratinocytes must be carefully planned to ensure that the requisite number of grafts are available on the scheduled operating day. The procedure we use is described in Navsaria *et al.*, Chapter 1.

(ii) The keratinocytes to be grafted must be just confluent otherwise they will be difficult to handle. It is widely accepted that it is best to use the graft within two days after the keratinocytes have reached confluence for optimal clinical results, although no clinical trial has been performed to establish this to date.

(iii) Keratinocytes are grown in tissue culture flasks rather than dishes in order to minimize the risk of contamination. On the day of grafting, the top of each flask is cut off by applying a hot soldering iron to the sides. Medium is removed and the cultures washed; at confluence no feeders remain. The enzyme Dispase (Sigma, 2 mg per ml in serum-free Dulbecco's modified Eagle's medium (DMEM)) is added to cover the culture and the culture is placed in an incubator at 37 °C.

(iv) Within an hour, the graft will start detaching from the edges of the flask and will curl inwards as the sheet shrinks towards the middle of the flask. When this starts happening, the backing dressing (Johnson and Johnson's nonadherent(N-A) dressing; Surfasoft or fine vaseline gauze) is placed on top of the graft. This will keep the graft evenly spread, provide support to the graft and preserve the orientation of the cell layers so that the basal cells are applied to the wound surface.

(v)   The dressing is stapled to the grafts with an operating stapling gun, or sutured with 5/0 silk.

(vi) The grafts are washed three times with PBSA, and very gently removed from the flask by holding the edges of the graft with a pair of sterile forceps. The graft is transferred to a sterile petri dish. Serum-free DME is added (enough to cover the graft) and the graft is then transported to the operating theatre. It is, of course, essential that the graft is prepared and transported under sterile conditions.

(vii) Grafting should be carried out as soon as possible. If a few hours delay are unavoidable, the grafts should be stored at 4 °C.

Preparation of the wound bed

A chronic granulating wound bed is a poor surface for keratinocyte grafts. A freshly excised bed is preferable but pretreatment of the bed with live dermis from a relative of the patient or with cadaver skin appears optimal and increases graft take dramatically. Thus, while awaiting grafting, whole allograft skin is used to dress the wound. At the time of grafting the cultured keratinocytes, the allografted epidermis is removed by shave excision or keratotome. An imprint of dermal remnants should be left.

Application to the patient

(i)   The surgeon should be instructed in the handling of the grafts and their orientation. Ideally, the technician who prepared the grafts should accompany them to theater.

(ii)   Once applied to the wound bed, the grafts must not be moved, as lateral shear will damage the germinative basal cells. To prevent slippage, the grafts are stapled to the patient.

(iii) Top gauze dressings are applied to the grafts, but antiseptics must not be added at this stage as they are highly toxic to keratinocytes (Tatnall, Leigh & Gibson, 1991).

(iv)   The patients are nursed on pressure avoidance beds, but movement is limited and only the top dressings can be damaged.

(v)   The patients are returned to theater at one week, and staples and backing dressings gently removed without trauma.

(vi)   The skin is very fragile and touch/pressure should be avoided. Graft take is very difficult to assess at one week as only slight translucence is seen in grafted areas.

(vii) Careful redressings are performed as required. Mild topical antiseptic preparations may be used after two weeks when a uniform opaque change in the graft makes areas of take possible to assess. The skin remains fragile and trauma, for example, application of sticking plaster, must be avoided.

(viii) Further grafting can take place as required.

12 Composite grafts containing Dermagraft™ as a dermal replacement

J. HANSBROUGH

When placed on a full-thickness wound, cultured sheets of pure keratinocytes lack a dermal component. The rete ridge interdigitation at the normal dermal–epidermal junction is thought to provide strength to the DEJ and increase resistance to shear forces. The dermis and associated matrix components are helpful for optimal wound healing. Reconstitution of the dermal–epidermal junction, and in particular the basement membrane, aids skin strength and durability.

In an effort to develop an effective cultured skin replacement, our group developed a composite skin replacement which employed cultured human keratinocytes (HK) and human fibroblasts (HF) combined with a dermal matrix. This material consisted of HK which were placed on the surface of a collagen–glycosaminoglycan (GAG) matrix which had been previously seeded with cultured HF. The collagen matrix may then function as a dermal template, allowing fibrovascular ingrowth from the wound bed similar to that seen with the Burke-Yannas collagen–GAG–silicone elastomer skin replacement (Yannas et al., 1980). In addition, we have shown that the inclusion of both HK and HF in the composite grafts results in improved basement membrane formation, increased production of laminin and type IV collagen in vitro, and thicker epithelial layers (Cooper et al., 1993).

In animal studies (Cooper et al., 1991, 1993) we found that these composite grafts achieved excellent 'take' on clean, full-thickness wounds. Immunofluorescent staining identified laminin and type IV collagen and electron microscopy confirmed the presence of basement membrane components by 10 days after grafting.

Seven patients with 43–97% TBSA burns were treated with autologous composite cultured grafts (Hansborough et al., 1989), which were applied to previously excised wounds, which were closed initially with cutaneous allograft. Overall 'take' of the composite grafts was approximately 50%. Complete basement membranes were found by 10 days of placement of the composite grafts. However, the collagen–GAG matrix appeared to be highly susceptible to bacterial and enzymatic attack, and applying these grafts to chronic wounds heavily colonized with bacteria is generally unsuccessful.

In addition, human keratinocytes are known to produce procollagenase, the metalloproteinase which initiates the degradation of native collagen.

In an effort to develop an improved dermal matrix, we have worked with Advanced Tissue Sciences, Inc. (La Jolla, CA) on developing a living dermal tissue replacement which is composed of HF cultured on a biodegradable, synthetic mesh made of polyglactin-910 (Vicryl^{TM}, Ethicon Inc., Somerville, New Jersey). The material has been termed Dermagraft^{TM}. Various surgical specialties have employed these fibers in the form of woven or knitted meshes, as they generate a relatively limited inflammatory reaction and are primarily biodegraded by hydrolysis rather than by enzymatic degradation.

Preparation of living dermal replacement tissue (Dermagraft^{TM})

To construct the living dermal replacement (Dermagraft^{TM}, Advanced Tissues Sciences, La Jolla, CA) (Hansbrough *et al.*, 1992; Cooper *et al.*, 1993; Hansbrough *et al.*, In press; Slivka *et al.*, 1991*a*, 1991*b* Landeen *et al.*, 1992), human dermal fibroblasts are isolated from neonatal foreskin obtained aseptically after circumcision. Epidermis and dermis are separated by incubation in 0.25% trypsin/0.2% EDTA for 1 to 2 h at 37 °C. Dermis is minced and digested with collagenase B, and the tissue digest is filtered through sterile gauze to remove debris. Fibroblasts are maintained in Dulbecco's modified Eagle's medium (DMEM), and are passaged at 80% to 90% confluence. Fibroblasts are removed from flasks and resuspended for seeding at a concentration of 4×10^6 cells/ml. Living dermal grafts are prepared by seeding 4×10^5 viable fibroblasts, with viability determined by trypan blue exclusion, in a minimum volume of DMEM onto each 4 cm^2 area of polyglactin-910 surgical mesh (Vicryl^{TM}, Ethicon Inc., Somerville, New Jersey). This mesh consists of knitted Vicryl^{TM} strands, with a pore size of 280×400 μm and a fabric weight of 1.50 oz/yd^2. The diameter of the Vicryl^{TM} fibers is 100 μm. The fibroblasts readily attach *in vitro* to the mesh fibers and become confluent in 2 to 3 weeks (at confluence all mesh openings are covered by cells and tissue matrix, and this can be assessed by inverted phase microscopy).

Cryopreservation of Dermagraft^{TM}

Living dermal grafts are cryopreserved after the fibroblasts reach confluence. The cryoprotectant solution consists of DMEM, 20% fetal bovine serum, and 10% glycerol. Grafts are frozen to -70 °C in a controlled-rate freezing unit (Gordin Ier Electronics, Roseville, Michigan) at a rate of 1 °C/min. Grafts are thawed in a 37 °C water bath for 5 min just prior to application of keratinocytes. A neutral red dye uptake assay is used to determine cell viabilities after thawing. Viabilities are greater than 80% of control cells (nonfrozen).

Application of cultured keratinocytes to Dermagraft™ surface

Human keratinocytes (HK) are isolated from split-thickness skin graft specimens. Skin is minced and 0.025% collagenase (Gibco, Gaithersburg, MD) + 10% v/v bovine pituitary extract is added. After 30–45 min in the collagenase solution, the epidermis is mechanically separated from the dermis. The HKs are then dissociated into a single cell suspension by treatment with 0.025% trypsin and 0.01% ethylenediaminetetracetic acid (EDTA) (both from Sigma Inc., St. Louis, MO) for an additional 5 min. Trypsin is neutralized with 10% fetal bovine serum. HKs are resuspended and expanded in number in 75 cm^2 polystyrene tissue culture flasks with KBM™ (Keratinocyte Basal Medium, Clonetics Inc., San Diego, CA) containing 0.15 mM calcium, increased amounts of selected amino acids, 10 ng/ml epidermal growth factor, 5 μg/ml insulin, 0.5 μg/ml hydrocortisone, 0.5% vol/vol bovine pituitary extract, and gentamicin surface (50 μg/ml)-amphotericin B (50 ng/ml) (Clonetics Inc., San Diego, CA).

When HK cultures are grown to confluence and ready for harvesting, trypsin/EDTA is added. Trypsinized cells are pelleted in KBM™, counted, and 1.4×10^8 keratinocytes are layered onto the surface of a 10×10 cm section of thawed Dermagraft™, in flat-bottom tissue culture vessels in supplemented KBM™ medium. HKs are grown to confluence on the Dermagraft™ surface over a 4 to 6 day period.

Application of composite grafts to wounds

Excess culture medium is removed from the petri dishes and the composite grafts are removed from the dishes. The Dermagraft™ possesses excellent structural stability so that the composite grafts may be transferred to the wound without further support. In addition, the grafts can be stapled or sutured to the wound bed because of the durability of the Dermagraft.™

Comments

As they grow in the Vicryl™ mesh, the fibroblasts secrete proteins and glycoproteins resulting in formation of an extracelluar matrix which fills the mesh interstices. Clinical trials are in progress to test Dermagraft™ as a dermal replacement beneath meshed skin grafts Hansbrough *et al.*, 1992. Composite grafts described above may allow wound closure with a bilayered skin substitute which is resistant to wound proteases. Application of these grafts to full-thickness wounds on athymic mice has been performed (Hansbrough *et al.*, 1993) is under way. Currently, 'take' and wound closure has occurred in approximately 50% of the wound areas which have been grafted. Histologic analysis reveals a well-differentiated epithelial layer and continuous staining for the basement membrane protein laminin at 20 days postplacement.

Because the Vicryl™ mesh has good structural characteristics, including high resistance to tearing (Cohen *et al.*, 1992), composite grafts composed of Dermagraft™ and cultured keratinocytes are easy to handle and transfer. Dermagraft™ holds staples and sutures well.

Part II

ASSAYS OF KERATINOCYTE ADHESION, PROLIFERATION AND GROWTH FACTOR PRODUCTION

This section contains techniques for assaying keratinocyte adhesion, proliferation and growth factor production. The adhesion assay measures attachment to extracellular matrix proteins and can be used to compare the adhesiveness of different subpopulations of keratinocytes in order to identify the cell surface receptors involved (see Hotchin and Watt, Chapter 13). Detailed protocols for assaying proliferation by determining the proportion of cells entering mitosis or the proportion of cells in S phase of the cell cycle are included (see Dover, Chapter 14), together with a simple technique for measuring cell number by DNA content (see Otto, Chapter 15). The pros and cons of a range of other proliferation assays are evaluated (see Dover, Chapter 14). The final part of this Section includes a bioassay for IL-6, which serves as a model for how to measure growth factor production of keratinocytes (see Dalley *et al.*, Chapter 16).

13   Extracellular matrix adhesion assays

N. HOTCHIN and F. M. WATT

Introduction

Interactions between keratinocytes and extracellular matrix (ECM) proteins, mediated by the integrin family of receptors, regulate many aspects of keratinocyte behavior (for review Watt & Hertle, 1994). The method which follows can be used to determine the proportion of keratinocytes within a given population that is capable of adhering to specific extracellular matrix proteins. It is also suitable for testing the ability of anti-integrin antibodies to inhibit adhesion.

Preparation of assay plates

(i)   Coat 96 well microtiter plates (Immulon 2, Dynatech) with 100 μl ECM protein per well. Typically, ECM concentrations of 10–100 μg/ml diluted in PBS are used. For each assay condition, use triplicate wells.

(ii) Incubate plates in a humidified atmosphere overnight at 4 °C.

(iii) Wash plates 3 times with PBS.

(iv) Incubate plates with 0.5 mg/ml heat-denatured bovine serum albumin (BSA) diluted in PBS for 60 min at 37 °C to prevent non-specific cell attachment. Denature BSA by heating a 10 mg/ml stock of BSA in PBS at 80°C for 3 min.

(v)   Wash 3 times with PBS.

(vi) Add 50 μl serum-free medium (we use FAD: 1 part Ham's F12, 3 parts DMEM and 1.8×10^{-4} M adenine) containing concentrations of antibodies or peptides for adhesion blocking/promotion experiments.

Labelling cells with ^{51}Cr (adapted from Brunner, Engers & Cerottini, 1976)

(i)   Harvest keratinocytes using trypsin/EDTA.

(ii)   Neutralize trypsin with 0.5 mg/ml soybean trypsin inhibitor (Sigma).

(iii) Wash cells twice with FAD and resuspend 10^6 cells in 300 μl FAD.

(iv)   Add 100 μCi ^{51}Cr (Amersham CJS1; 1 mCi/ml, made isotonic by addition of 1/10th volume 10 × PBS).

(v)   Incubate for 45 min at 37 °C, gently swirling cells every 15 min.

(vi)   Wash twice in FAD and resuspend cells at 2×10^5/ml.

Keratinocyte methods by Irene Leigh and Fiona Watt
© Cambridge University Press, 1994, pp. 73–74

(vii)   Add 50 μl (i.e. 10^4 cells) per well of microtiter plate.
(viii) Also prepare a standard curve, in triplicate, using serial dilutions from 10^4 to 10^2 cells.

Quantitation of cell adhesion

(i)   After incubation for the desired length of time at 37 °C in a humidified atmosphere, remove unbound cells by rapidly inverting the microtiter plate and flicking out the liquid.
(ii)   Wash 3 times with PBS.
(iii) Release chromium from bound cells by adding 75 μl 0.1M NaOH, 1% SDS, 2% NaHCO$_3$.
(iv)   Leave for 10 min at room temperature before transferring lysate to counting vials. Cotton-tip applicators work well for this purpose.
(v)   Determine cpm in lysates by counting in a γ-counter.
(vi)   Using the standard curve convert cpm to number of cells bound per well.

Notes
Standard curves should be prepared for each cell type or treatment. For example, if comparing adhesion of keratinocytes before and after suspension in methyl cellulose (see Watt, Chapter 20) standard curves for both the starting population and for suspended cells must be prepared.

For each experimental condition, nonspecific adhesion to wells coated with BSA alone should be determined and subtracted from adhesion to ECM.

As an alternative to serum-free FAD medium, Tris buffered saline, pH 7.4, can be used.

Two alternatives to chromium labelling cells are to score the number of attached cells visually or to quantitate on the basis of hexosaminidase activity. For these methods, and for examples of keratinocyte adhesion on different concentrations of ECM proteins, and the effects of different incubation times, competing peptides and anti-integrin antibodies, see Adams and Watt (1990, 1991).

14   Assays of keratinocyte proliferation

R. DOVER

Introduction

It is often necessary to assess the effects of exogenous growth factors or other stimuli on keratinocyte proliferation. Keratinocytes, however, are heterogeneous: some cells may be actively proliferating (i.e. in cycle), others may be quiescent, and others may have initiated terminal differentiation and thus be incapable of further proliferation. An increase in cell number may reflect an increase in the proportion of keratinocytes that are in cycle or a decrease in cell cycle time. Some stimuli may exert their effects within a few minutes or hours; for others, continuous exposure over a number of days may be necessary to see an effect.

In view of this complexity, I have outlined a range of methods that can be used to analyse keratinocyte proliferation. Two basic techniques are described in detail. The first is the stathmokinetic method, in which cells are arrested in metaphase by vincristine, in order to determine the proportion of cells entering mitosis over a given period of time. The second method is determination of the labelling index of a population of keratinocytes: the proportion of cells in S phase of the cell cycle is measured by labelling with ^{3}H thymidine (3HTdR). Both techniques have been successfully applied to cultures of human keratinocytes (see for example, Dover & Potten, 1983).

Stathmokinetic method

Materials

- Cells on coverslips
- Phosphate buffered saline, PBS
- Culture medium
- Vincristine sulphate
- Syringe and needle
- Micropipette 1–20 μl and sterile tips
- Forceps
- Fixative
- Stain
- Mounting medium

Keratinocyte methods by Irene Leigh and Fiona Watt

The first stage of this procedure is to determine the optimum vincristine concentration and exposure time.

(i)   Prepare sterile solution of a dose range of vincristine sulphate in PBS or culture medium. A suggested starting point would be to use doses of 5, 4, 3, 2, 1.5, 0.75, 0.5, 0.25, 0.1 μg/ml. It is best to prepare and dilute only what is needed and discard any left over.

(ii)   Using at least two coverslips for each dose, treat the cells with the vincristine sulphate for 2.5 h prior to fixation.

(iii) Stain the cells and mount and examine under a microscope.

(iv)   Counts of metaphases are then made and plotted against dose. The optimal dose which produces the maximum number of metaphases should be selected. If results are similar over a range, the lowest dose should be chosen to avoid toxicity. It may be necessary to repeat the process over a narrower dose range to be sure of choosing the least toxic, most potent dose. It is essential to note the presence of any anaphases or telophases in any of the samples as this indicates incomplete arrest and that higher doses should be used.

(v)   With the dose determined as above, a time course study must now be performed. Treat at least 2 coverslips per time point. Treat the cells with vincristine for a range of times up to 5 h at 30 min intervals.

(vi)   Fix, stain, mount and count.

(vii) Plot the results.

It should be clear that at the longest incubation times the curve of number of metaphases versus cell number has reached a plateau and the mitotic figures in the samples are degenerating and condensed. The early part of the curve may also be nonlinear. By inspection of the curve, the safe range of incubation times can be determined, where metaphase arrest is linear.

Having determined the optimal conditions for vincristine treatment, the metaphase index can be measured in control and experimental samples, as follows.

(i)   Cells on coverslips are set up and treated with the determined optimal dose.

(ii)   Multiple samples, at least duplicates, are then sampled at intervals over the safe collection period determined as above. Typically, this might be every 30 min over 3–4 h.

(iii) The cells are then prepared for microscopy and counted as usual.

(iv)   A graph of metaphase index versus time can be plotted and a least squares fit made to the line. Ideally, 95% confidence intervals should also be determined as this gives a clearer view of the proliferative rate. The slope of the fitted line will give the rate of entry into mitosis or the rate at which cells are born into the population.

Labelling index

Materials

- Cells on coverslips
- Tritiated thymidine
- Gloves
- Micropipette and tips
- Prewarmed medium
- PBS cooled to 4 °C
- Fixative
- Stain
- Mountant

(i) The worker must be familiar with the safe handling and disposal of radioisotopes, and the local rules within their institution governing the use and storage of the material. For most kinetic studies, the doses of radioactivity used are small but, as the labelling agent is most often thymidine, it is incorporated directly into the DNA. The most common isotope used is tritium. This produces β-particles with a short path length in biological fluids and so most damage would be local, i.e. within the DNA itself. It must therefore be borne in mind that, despite the low doses of radioactivity involved, any accidentally incorporated label is targeted with high efficiency to the genetic material and that tritium has a long half-life in human terms (about 12 years). The use of simple good laboratory practice can eliminate the risk of contamination: disposable gloves should be worn and the work performed and solutions handled in a tray in order to contain any spillages.

(ii) Tritiated thymidine is available at a number of specific activities. The specific activity is a measure of the proportion of the thymidine that is actually labelled, the higher the specific activity, the higher the ratio of tritiated molecules within the solution. In most situations a specific activity of 5 Ci/mmol (185 GBq/mmol) is adequate; higher activities are more expensive to purchase. Tritiated thymidine can dissociate with time, leading to a decrease in labelling activity which can be as high as 1% per month. It is thus important to record the delivery date of the solution and refer to the manufacturer's recommended storage conditions and times. For most cells a dose of 1 μCi/ml (37 kBq) will give good results, and it is simple to remember that this corresponds to a 1:1000 dilution from a stock of 1 mCi/ml (37 MBq/ml) that is commercially available.

(iii)   The thymidine should be diluted to give a working stock. If there are a large number of samples to label, it is most efficient to prepare a stock solution of labelled culture medium and add this to the cells. The medium should be prewarmed to the incubation temperature of the cells; adding cold medium can impair cellular uptake of the label and give false results. For small numbers of samples, the thymidine can be dispersed using a micropipette, adding as small a volume to the culture medium as possible to avoid diluting effects on the osmolality of the medium. Using 1 μl/ml of a stock of 1 mCi/ml will give a final concentration of 1 μCi/ml.

(iv)   To obtain a simple labelling index, the cells would normally be exposed to the isotope for a period of 1 h. It is important to maintain accurate timing. If large numbers of cultures are to be treated, it can take some time to process them all. A 6 min delay between first and last samples is a 10% error on top of any other systematic errors. It is preferable to use smaller groups of cultures and stagger the treatments than to introduce such errors. For some experiments continuous labelling may be required. This is performed in exactly the same manner, but the coverslips are sampled at intervals over a given time period.

(v)   After the desired labelled period, the radioactive medium is removed and the cultures washed twice with cold PBS (4 °C), the medium and washes should be disposed of according to local safety rules for disposal of radioactivity.

(vi)   The cultures are fixed. Mercury-containing fixatives may remove DNA from the cells and are not suitable, neither are fixatives which give a compact nucleus, such as Bouin's. Unless there is a good reason for choosing a different fixative (e.g. combining autoradiography with immunocytochemistry), fixation in 70% ethanol or 1:1 acetone: methanol should produce good results. If these fixatives are used, the slides can be air dried; in other cases (for example, para-formaldehyde fixation) it is preferable to hydrate, wash, give a final distilled water rinse and then air dry.

(vii)   At this stage, fixed cells can be stored in a dust-free environment, but it is often easier to mount them on microscope slides, cell side uppermost. The coverslip can be stuck down with mounting medium but this normally requires several hours to harden. A better method is to use a glass–glass adhesive which bonds under the influence of daylight. Glassbond^{TM} (Loctite UK Ltd) is such an adhesive. Coverslips can be taken direct from water and the underside wiped clean of excess fluid. Traces of water do not appear to inhibit the bonding, and it can be easier to manipulate the coverslips whilst still wet. A drop of Glassbond is placed on the microscope slide and the coverslip picked up with fine forceps and placed (cell slide upwards) onto the adhesive. Practice is needed to ensure just sufficient adhesive is used; any excess should be wiped off immediately. Multiple coverslips can be mounted on one slide if required.

The slides are placed in a dust-free sunlit place. In strong sunlight, bonding occurs within seconds. On overcast days, it can take 10–15 min. Bonding can be tested by pushing on the edge of the coverslip gently with fine forceps; once bonded, the coverslip will not move. In an emergency, a UV source can be used to induce bonding; face protection should be worn and the bonding time will vary depending on the power of the UV source.

(viii) Once bonded, the slides can be stored in a dust-free environment prior to further processing. Slides would then be autoradiographed, stained and counted as described below.

(ix) The work area should be monitored by swabbing the area and counting the swabs in a scintillation counter. Records must be kept of the amount of radioisotope used and the route of disposal.

Autoradiography

Preparation

Dipping jars
In practice, any vessel would suffice but it is better to use a specially designed container to minimize the amount of emulsion used as it is expensive and not reusable. A cut-down measuring cylinder is one simple form of a dipping jar.

Emulsion
The most commonly used emulsion is a dipping film. An alternative is a stripping film but this is more expensive, difficult to handle and more time consuming to use.

The dipping film emulsion comes from some suppliers as a solid mass in a small jar; it can all be melted at once or else a small amount can be cut out, taking extreme care not to physically damage the emulsion as this can lead to high backgrounds during autoradiography. Other companies supply the emulsion as shreds of gelatinous material which can be more easily dispersed. Emulsions can be cooled, remelted and used again but this is not recommended as it is very simple to damage the emulsion and increase the background. In general, the minimum amount of emulsion required should be melted and any that is left over should be discarded. Whilst this may seem wasteful, it is cheaper than having to repeat an entire experiment if the background is too high.

The quality of the emulsion varies with the nature of the radioisotope used and the sensitivity required; refer to the manufacturers' data sheet when selecting an emulsion.

Controls

For each new procedure, a set of control slides is required, since stains, fixatives or cell products can potentially interfere with the emulsion, reducing or enhancing the signal. Cells not exposed to radioactive precursor, but treated identically in every other respect, should be prepared. Positive controls to test the function of the emulsion are also required.

General

Cleanliness is essential; dust will cause major problems. Temperature control is also important to prevent physical stress to the emulsion. Only nonmetallic implements should be used (i.e. plastic slide racks, glass rods for mixing), as the reagents react with metal.

Autoradiographic method: dipping film

Materials

- 40 °C waterbath
- Nuclear emulsion (e.g. Ilford K2)
- 25 ml measuring cylinder
- Clean glass microscope slides
- Dipping jar
- Distilled water
- Glycerol (optional)
- Glass rods
- Lightproof boxes
- Silica gel in cloth sachets
- Tape
- Safelight
- Lint-free cloth or tissue paper
- Cooled tray or metal sheet on ice
- Darkroom with filters suitable for the emulsion in use, see manufacturer's data sheet
- Experimental slides
- Positive and negative chemography controls

Preparation

(i)   In the darkroom with the main illumination on, assemble all the equipment and allow the water bath and the tray to reach the required temperature. A piece of flat metal placed on a thin layer of ice can be substituted if a cooled tray is not available.

(ii)   Measure 12 ml of distilled water into the measuring cylinder and place in waterbath. Alternatively use 11.75 ml distilled water and 0.25 ml glycerol. Place the empty dipping jar in the waterbath to warm up.

(iii)   Although the slides can be dipped when dry, emulsion coverage is more even if they are first rehydrated. Use double distilled water and ensure that the slides are all aligned in the same direction, so that the user can tell which side of the slide has the mounted cells even in the dark.

(iv)   Turn off the main illumination in the darkroom and work with only the orange safelights on. It may take a few minutes for your eyes to adjust.

(v)   Remove the emulsion from its packaging and add emulsion to the 12 ml of prewarmed water (or water/glycerol mix) in the dipping jar. Add until the meniscus reaches the 24 ml mark. Reseal the unused emulsion and ensure it is light-proof.

(vi)   When the emulsion is melted, after 5–10 min, stir gently with a glass rod, avoid producing bubbles. It is essential that the emulsion is evenly mixed. Pour the emulsion into the dipping jar.

(vii)   Take a blank slide and place it into the emulsion, withdraw slowly at a constant rate to ensure that the thickness of the emulsion is the same over the entire surface of the slide. The first few slides may have bubbles on them; if so, continue using blank slides to remove them. These slides can be discarded, but could serve as auto-radiographic controls.

(viii) When the emulsion is bubble free, the experimental slides can be dipped.

(ix)   Take the first slide and remove any excess water. Dip the slide in the emulsion and then withdraw slowly and evenly. Use a damp cloth or tissue to wipe excess emulsion off the back of the slide. Place the slide with cells up on the cold tray.

(x)   Continue dipping slides and collecting on a cold surface. Dip the positive and negative chemography controls (see (xiv) below).

(xi)   When all the slides are dipped, remove the tray from the ice to prevent frosting on the slides.

(xii)   The slides must dry in a dust-free and light-free environment for a minimum of 3 h. The ambient temperature in the darkroom should not be too high, or this will increase the background. If possible, leave the slides overnight to dry in the darkroom with all lights off. It is wise to leave a large note on the door to ensure no one enters and turns on the light!

(xiii) Remove the equipment from the darkroom and wash thoroughly. Unused emulsion may be collected for silver extraction and recycling if used in sufficiently large quantities.

(xiv) When the slides are dry, they should be boxed in light-proof containers and labelled with the date of preparation and experiment details. Each box should contain a sachet of silica gel as a drying agent, this can be placed at one end of a slide box and segregated from the experimental slides by a blank to prevent mechanical damage to the emulsion.

When the experimental slides are sealed in a light-proof container, a positive control blank and the negative chemography control slide can be prepared. The negative chemography control is a slide of cells that have not been exposed to radioactive label; this slide is dipped in emulsion. The positive control is a blank slide dipped in emulsion; simply turning on the main light for 1 s should be enough to fog the emulsion on both the slides of the positive control. Box-up the control slides as for the experimental slides.

(xv) Place the slide boxes in a refrigerator, avoiding positions close to the compressor which may emit small amounts of electromagnetic radiation. Storage of autoradiographs near to any other sources of radiation should also be avoided.

Development

Materials

- Dipped slides
- Positive and negative chemography control slides
- Autoradiography control slides
- Developer (Kodak D19 or Ilford ID11)
- Distilled water or 1% acetic acid
- Commercial photographic fixer or sodium thiosulphate 20–40% solution w/v
- Staining dishes and nonmetallic slide racks
- Large tray to hold slide racks
- Ice
- Thermometer
- Timer

(i) Fill one slide dish with undiluted developer. Stand in tray of water and ice to adjust temperature to 18 °C.

(ii) Fill second dish with distilled water or 1% acetic acid in distilled water

(iii) Fill third dish with Fixer prepared according to manufacturer's instruction for film. Alternatively, use a 20–40% solution of sodium thiosulphate.

(iv) When the solutions are at 18 °C, switch out the main light and, using only the safelights, unbox the slides and place them in a slide rack.

(v)   Put the slides in the developer; start the timer. Tap the slide rack to dislodge any bubbles. Agitate the slides *gently* every 30 s. Continue development for 5 min.

(vi)   Remove the slide rack and briefly drain excess developer off. Transfer the slides to the water or dilute acetic acid and agitate for 30 s.

(vii) Drain the slides and transfer to the fixer. Fix for the time suggested by the manufacturer. If using thiosulphate, pretest it by dipping a small piece of photographic film into the solution. It will gradually go clear. Fix for three times the clearing time.

(viii) The lights may now be switched on. The slides should now be washed in cold running tap water for a minimum of 5 min.

(ix)   The slides are stained with the required histological stain and mounted. As the slides already have coverslip (with the cells) stuck to them they are rather thick. This means that it is best to use either an equal sized coverslip to just cover the sample, or if multiple cell-bearing coverslips are used a large oversized coverslip is used to cover them all. The slides can now be examined and scored.

(x)   The positive control slide, which was deliberately exposed to light should be totally black. The negative control, a blank slide dipped in emulsion but not exposed to light, should not be black! It will have a background level of grains on it caused by cosmic radiation, dust contamination and drying and stretching effects.

Critique of methods for assessing keratinocyte proliferation

Data on keratinocyte proliferation have been obtained using a wide range of techniques. The data are discussed in detail by Dover (1994; *Keratinocyte Handbook*). Brief critiques of individual methods are given here.

Cell number/growth curve

Equal numbers of cells are seeded per dish in the presence or absence of test substances and, at intervals, duplicate or triplicate dishes are harvested and the number of cells per dish determined with a hemocytometer or Coulter Counter.

Pros

A simple and direct method which will estimate number of cells per dish.

Cons

Cells shed into the medium should also be counted. It is theoretically possible that the rates of cell production and cell detachment increase simultaneously, so that, although the population of cells produced is increased, the number of cells per dish remains constant. This method will not detect changes in the types of cells within the culture, e.g. the proportion of proliferative vs differentiated keratinoctyes.

Comment
A useful indicator of possible changes as it is simple to perform.

Protein/enzyme content

Instead of direct measurement of cell number, assays of total protein or housekeeping enzymes can be performed. Cell number is then inferred from standard curves of cell number versus total protein or enzyme assayed. Examples of appropriate enzymes are hexosaminidase, which has been used in keratinocyte adhesion assays (Adams & Watt 1990), and 3-(4,5-dimethylthiazol-2-yl)-2,5-diphenyl tetrazolium bromide (MTT) which has been used in bioassays of growth factors (see Dalley *et al.*, Chapter 16).

Pros
Simple to perform; rapid; multiple samples can be assayed easily.

Cons
Potentially very misleading because the assays are so indirect. As keratinocytes are a heterogeneous population, enzyme levels and protein content per cell can vary with stage of differentiation. Furthermore, any drug treatment may selectively affect the levels of the enzyme without affecting proliferation. Similarly, changes in protein content per cell may occur with experimental treatment, again leading to false indications of cell number.

Comment
The methods should only be used after extensive testing to ensure that any changes detected truly reflect changes in cell number.

Surface area

The surface area of a keratinocyte colony can be used as a measure of the number of cells it contains (see, for example, Barrandon & Green 1985).

Pros
This is perhaps the simplest assay of keratinocyte proliferation, particularly with the availability of automated image analysis techniques.

Cons
Potentially very misleading. Time lapse observations of small keratinocyte

colonies show that the area of colonies can change with time without a change in cell number, because colonies are motile and the cells at the trailing edge of a colony can become temporarily extended. Motility of colonies can lead to differences in area of at least 30% without any change in cell number. Similarly, measurement of colony area does not account for any changes in the number of cell layers or for any changes in cell size or spreading.

Comment

This method should be avoided unless validated by cell counts for every experimental condition to be tested.

^{3}H-TdR cpm/μg DNA

As a measure of new DNA synthesis, ^{3}H thymidine incorporation is expressed relative to the total DNA content of the cell population.

Pros

Replicates are simple to perform and the assay is rapid.

Cons

Fraught with potential sources of error. Changes in thymidine kinase, the enzyme required for thymidine uptake, can be induced by drug treatment. Thus changes in incorporation of ^{3}H-TdR do not necessarily reflect proliferative changes. Changes in the rate of incorporation (i.e. changes in S-phase duration) are also possible, thus invalidating the method. There is considerable evidence that keratinocytes handle thymidine in an unusual manner and that its incorporation does not necessarily reflect proliferative rates (see Dover, *Keratinocyte Handbook*).

Comment

This technique is unreliable at best and should be avoided if at all possible

Mitotic index (MI)

The proportion of cells in mitosis is assessed microscopically

Pros

Simple to carry out. No special equipment required.

Cons

It can be technically difficult to score samples as some phases of mitosis (e.g. prophase) are hard to recognize. One solution is to count only metaphases as these are easy to see; however, it is not known if the ratio of time spent in each phase of mitosis varies in cultured keratinocytes. Changes in the overall duration of mitosis (up to 100%), have been reported for human keratinocyte cultures. Thus, to count a static mitotic or metaphase index, which assumes a constant duration, may lead to errors of at least 100%. The second problem is how to express the results: should they be mitoses as a proportion of all keratinocytes, all nucleated keratinocytes or all basal keratinocytes?

Comment
Although, *in vivo*, the MI can be a powerful and accurate estimator of proliferation, it is probably not as reliable in keratinocyte cultures.

Labelling index (LI)

The percentage of cells in S-phase is determined by measuring ^{3}H-TdR or BrdU (bromodeoxyuridine) incorporation into DNA. Incorporation can be assessed microscopically or, in the case of BrdU, by flow cytometry.

Pros
Simple to apply. Flow cytometry is rapid to perform.

Cons
The problems concerning thymidine metabolism in keratinocytes (see above) apply to both ^{3}H-TdR and BrdU uptake, since both substances are taken up by a common pathway. BrdU is toxic and may, in itself, perturb the cell cycle; thus it should generally be avoided other than as a tertiary label (i.e. when the cells are fixed immediately after exposure). The question of denominators raised for MI also applies to this technique. The results are subject to error from changes in S-phase duration.

Comment
Despite the problems, LI can be measured quickly, particularly with BrdU, and is a useful indicator of gross changes. However, it would be wrong to infer that changes in proliferation have occurred from LI evidence alone.

Total DNA content

For protocol see Otto, Chapter 15.

Pros
DNA is the single constituent of cells which *a priori* must reflect the number of cells in a population. Thus, the assay of putative growth factors of inhibitors is most reliable when such an assay is undertaken. DNA assays can be quick, specific, and highly reproducible, and may even be automated given suitable equipment.

Cons

Since keratinocytes must be solubilized in order to determine DNA content, no positional data can be obtained; as it is not possible to determine whether a particular subset of cells are responding to an agent of interest. For normal cells, aneuploidy or polyploidy should not pose a problem but should be considered when tumor-derived keratinocytes are used.

Comment

A powerful and potentially very accurate method for estimating cell number.

Strathmokinetic method (metaphase arrest)

See detailed protocol above.

Pros

A direct assessment of the rate at which cells are being produced. Sensitive and simple to apply.

Cons

A suitable arrest agent (such as vincristine) must be tested first for dose response and toxicity.

Comment

A direct and sensitive method to detect cell production. In the absence of changes in cell death or loss, this will reflect the 'growth' of the culture as a whole.

Percentage labelled mitoses (PLM)/fraction labelled mitoses (FLM)

Cells in S phase are pulse labelled with ^{3}HTdR, and the proportion of mitoses that have ^{3}HTdR-labelled DNA is determined as a function of time of labelling (see Dover & Potten, 1983).

Pros

Powerful: reveals phase durations and cell cycle length.

Cons

Time-consuming and technically demanding, involves considerable micro-scopy.

Comment

Potentially very informative, but this technique is not for the faint-hearted as many samples must be prepared and scored over an extended period of time.

15 Determination of cell number by DNA content

W. OTTO

A fluorimetric DNA assay for cultured keratinocytes has been described (Rao & Otto, 1992) which combines the DNA specificity of Hoechst 33258, with the high sensitivity and rapid analysis provided by a microtitre plate-reader such as a Fluoroskan II (Labsystems). Liquid handling is kept to a minimum. This assay is useful for the study of putative growth factors or their inhibitors since it can discriminate ± 500 normal diploid human cells, while after the growth period plates may be stored frozen pending batch analysis, thereby allowing accurate time-series experiments to be undertaken from low cell number to confluence. The use of the 96-well format ensures that the quantities of expensive peptides or other agents affecting growth may be kept to a minimum, and yet enough replicates can be prepared for statistical rigor. The technique has advantages over other microtiter assays in that the buffer used to dissolve the keratinocytes is based on SSC, which inhibits DNases. Cornifying cells are dispersed by including urea in the SSC. The assay is fully compatible with the simultaneous use of tritiated thymidine given as a terminal pulse (1–6 h). Aliquots of the assayed cultures may be scintillation counted, allowing a useful check on salvage pathway incorporation into DNA. A brief outline of the assay is given below, while full details are available elsewhere (Rao and Otto, 1992). Experimenters are reminded that Hoechst dyes intercalate DNA and are mutagenic, therefore adequate safety procedures such as wearing gloves and avoiding aerosol production should be followed).

Simplified assay procedure

(i) Plate cells and treat as appropriate for 'n' days. The maximum number of cells per well that can be assayed is 10^5.

(ii) Aspirate medium from each well by gentle suction, or invert onto absorbent paper towels (if radioisotope-free).

(iii) Gently rinse out each well twice with $100\,\mu l$ PBS-A and aspirate. At this stage plates may be stored frozen for later analysis.

(iv) To each well add $100\,\mu l$ of 0.04% SDS diluted in $1 \times$ SSC with 8 M urea.

(v) Incubate plate at 37 °C for 1 h with occasional swirling.

(vi)   Add 100 μl of 1.0 μg/ml of Hoechst 33258 in 1 × SSC.

(vii) Transfer 100 μl aliquots to a white 96-well plate for maximum fluorescence sensitivity.

(viii) Read fluorescence of each well at excitation λ 355 nm, emission λ 460 nm, comparing with DNA solutions or cell 'standards' treated the same way. If assaying for radioisotope incorporation, further aliquots may be scintillation counted, or the whole well contents absorbed onto filters using a 'cell harvester' device. The filters are then read via a scintillant-free collimator system (i.e. Berthold 2000) or cut out and scintillation counted.

Solutions

SCC comprises 0.154 M NaCl, 0.015 M Na citrate pH 7.0 and may be made as a 20× stock. SDS is conveniently made as a 10% stock solution in high grade water (i.e. MilliQ or equivalent). Hoechst 33258 stock solution (1 mg/ml in SSC) is stable for several months when stored frozen at −20 °C. It is best to make fresh working solutions for each assay. Minimize exposure to bright light by covering solution containers with aluminium foil.

16   Growth factor assays

A. DALLEY, G. HOWELLS and I. A. McKAY

Introduction

Keratinocytes produce and respond to a wide range of growth factors
(McKay & Leigh, 1991). Here we describe several methods for measuring
production of growth factors by keratinocytes. A specific bioassay of
interleukin-6 (IL-6) production is used to illustrate some of the pitfalls to
be avoided when working with keratinocytes.

Methods for detecting growth factor production

Expression of growth factors can be detected by measuring mRNA levels,
by use of specific monoclonal antibodies, or by specific bioassays.

mRNA

Extraction and northern analysis of mRNA

Extraction of skin for detection of specific mRNA species has proven
problematical in the past. In looking for growth factor mRNA, it is
important to bear in mind that the turnover of some RNAs is very rapid
and so it is essential to rapidly freeze and/or extract skin samples before
the mRNA is degraded (Longley *et al.*, 1991). Moreover, the mere act of
taking a biopsy can cause induction of expression of specific mRNAs: we
have found that keratin gene expression can be induced during the time
taken for the biopsy to travel from operating theatre to laboratory. It is
worth bearing in mind that extracts of whole skin, even split-thickness skin,
will inevitably contain RNA from cells other than keratinocytes and that
specific expression by keratinocytes can only be demonstrated by *in situ*
hybridization using specific polynucleotide probes on sections of skin (see
below).

  mRNA can be extracted from cultured keratinocytes using any of the
standard procedures (Sambrook, Fritsch & Maniatis, 1989) and then
analyzed by northern blotting. However caution must again be exercised
over interpretation of results. Cultured keratinocytes show expression of
novel RNA species when compared with intact skin (Kopan & Fuchs,
1989). This presumably reflects their impaired ability to differentiate, their
increased rate of cell division and increased growth fraction. It is not safe
therefore to assume that mRNA expression detected *in vitro* reflects growth

Keratinocyte methods by Irene Leigh and Fiona Watt
© Cambridge University Press, 1994, pp. 91–96

factor production *in vivo*. Note also that variation in the level of growth factor mRNAs can reflect both differences in transcription rates and differences in RNA stability.

RT-PCR

Using reverse transcription coupled to the powerful amplification provided by the PCR technique it is possible to detect even very low levels of specific mRNAs. Protocols for the use of this technique in keratinocytes have been described by Longley *et al.*, 1991 and Van Zoelen *et al.*, 1993. It is important to note that this technique is at best semiquantitative; a control RNA of known concentration should be included in each PCR run.

In situ hybridization

mRNA expression can be detected in sections of skin using cDNA or cRNA probes labelled either radioactively or nonradioactively. This procedure has been used to detect expression of a variety of growth factors in skin, including TGFα (Turbitt *et al.*, 1990), IL-6 (Grossman *et al.*, 1989) and TGFβ1 (Akhurst, 1993) and protocols for its use in skin can be found in these references (see also Fisher, this volume).

Protein

Immunocytochemistry

Immunocytochemistry with specific antibodies has been used to demonstrate the presence of several growth factors in human skin, including TGFα (Gottlieb *et al.*, 1988; Elder *et al.*, 1989) and IL-6 (Yoshizaki *et al.*, 1990). Many proteins are sensitive to fixation, and some antibodies will only work on frozen sections; it is therefore necessary to identify the optimal conditions for staining cells or tissue sections with antibodies to growth factors. The specificity of this technique depends on the ability of the antibody to distinguish between different forms of the same factor and between different factors which exhibit some homology. Use of peptide immunogens which represent unique sequences in the factor can improve specificity (Harlow & Lane, 1988).

Immunoprecipitation, western blotting, RIA, SPA and ELISAs.

Immunoprecipitation, western blotting, radio-immunoassays (RIAs), scintillation proximity assays (SPAs) and enzyme-linked immunosorbent assays (ELISAs) all rely on antibodies to detect growth factors in samples from sera, cell supernatants and cell extracts (Capper, 1993). Caution must be exercised in interpreting the data from assays of serum and supernatants as they can be subject to interference by serum binding proteins, soluble receptors and nonspecific cross-reactivity (McKay, 1993).

Activity

Specific bioassays

Some cells depend on specific growth factors for their continued proliferation. This is particularly true of a number of hemopoietic cell lines, such as the murine B9 hybridoma cell line. The B9 line has been extensively characterized; it is unresponsive to a wide range of cytokines, but does respond to murine and human IL-6 (Aarden *et al.*, 1987). In our hands, B9 cells will not proliferate in medium supplemented with 10% FCS unless the medium is further supplemented with 20 units/ml IL-6. This absolute dependence on IL-6 for cell division is the basis of the specific bioassay. If supernatants from keratinocytes support the growth of B9 cells it can be inferred that the keratinocytes are secreting IL-6. Taken in conjunction with information derived from immuno-assays and RNA analysis, specific bioassays can confirm that a detected factor exhibits biological activity.

Growth factor bioassays may measure cell proliferation through ^{3}H-thymidine incorporation into DNA, through bromodeoxyuridine incorporation or through increase in cell number (See Dover, Chapter 14; Otto, Chapter 15). The growth assay described below is based on the ability of live cells to convert a yellow substrate, 3-(4,5-dimethylthiazol-2-yl)-2,5-diphenyl tetrazolium bromide (MTT) into a blue formazan product.

Each assay of cell growth can be adapted to a number of cell types but it is important to note that a correlation between the variable assayed (MTT conversion in this case) and cell number must be established for each assay.

Keratinocyte growth conditions for bioassay of factor production

Choice of medium

Any medium used to culture keratinocytes must be free of the growth factor to be assayed and must support the keratinocytes for long enough to allow production of detectable levels of factor. This may require some compromise as the factors which support keratinocyte growth and/or differentiation may well induce the production of factor to be assayed. The medium used for assay of IL-6 production by keratinocytes in the protocol below contains both hydrocortisone, which down-regulates IL-6 production, and epidermal growth factor (EGF), which can stimulate it.

Use of serum

In order to detect production of a particular growth factor in medium containing serum, it is necessary first to establish whether the serum contains factor(s) which might interfere with the assay. In each assay, a control of medium alone should be included. In the assay described here, we were unable to detect any effect of 10% FCS on B9 cell proliferation.

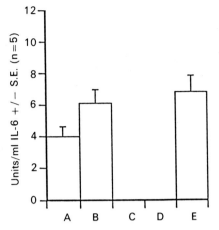

Fig. 16.1. Levels of interleukin-6 (IL-6) produced by human keratinocytes *in vitro*. A: SVK14 cells on collagen, B: SVK14 cells on plastic, C: HT-29 cells on collagen, D: HT-29 cells on plastic, E: Human foreskin keratinocytes on collagen, in the absence of feeder cells.

Care should also be taken to exclude the presence of soluble binding proteins or other factors in serum which might interfere with the assay. DTT and iodoacetamide treatment of serum removes most, if not all, interfering factors. A protocol for the production of DTT-treated serum supplemented with selenite and transferrin can be found in Van Zoelen *et al.*, 1993.

Use of feeders

If keratinocytes are cultured according to the method of Rheinwald and Green, then the production of factors by feeder cells must be considered. For example we found that the 3T3 feeder layer, produces 60 units/ml of murine IL-6 (cf Fig. 16.1). Moreover, the 3T3 cells continue to produce high levels of IL-6 after irradiation (Dalley & McKay, unpublished results). There are several ways in which this problem can be circumvented. Use of growth factor-defined media (see Mitra and Nickoloff, Chapter 4; Morris, Chapter 6) can remove the requirement for feeders as can plating keratinocytes at high density or on collagen–coated plates; alternatively, the feeders can be removed prior to the assay.

Collagen-coated plates

Using the SV40 transformed human keratinocyte cell line,SVK14 (Taylor-Papadimitriou *et al.*, 1982), we have found that plating on collagen-coated dishes does not affect the production of IL-6 (Fig. 16.1). Moreover, collagen does not induce production of IL-6 by the HT-29 human colon carcinoma cell line which serves as the negative control in this assay. Foreskin keratinocytes (FsK) produce similar levels of IL-6 when plated on collagen or on uncoated plastic, although there is greater variation between samples on collagen (Fig. 16.1).

Collection and storage of supernatants

In this assay, the supernatant is simply harvested and centrifuged to remove cellular debris. The clarified supernatants can then be stored at $-20\,^{\circ}$C until the B9 cells are ready for use. When other factors are to be assayed, it is important to determine their susceptibility to freeze–thawing. It may also be necessary to store the supernatants in siliconized tubes to prevent loss of factor through adhesion to plastic. If supernatants must be stored at $4\,^{\circ}$C do not add azide as preservative since this will affect the cells used in the bioassay. Filtering may also result in loss of factor. In the case of labile factors, it is best to perform the assay as soon after collecting the supernatants as possible.

Changes in cell number during incubation period

For the B9 assay, supernatants are harvested from the keratinocytes after 72 h. During this time, the keratinocytes may have divided, and the final cell number at harvest time should therefore be determined. In the assay presented here, there was no significant difference between the cell numbers for SVK14 and FsK. However, for comparisons between assays and between groups performing the same assay, it is best to record results as units/ml/fixed number of cells.

B9 bioassay of IL-6 secreted by human keratinocytes

Culture of human keratinocytes

Early passage foreskin keratinocytes (FsK) are maintained in a medium known as RM + : 3 parts Dulbecco's Modified Eagle's Medium (DMEM) and 1 part Hams F_{12} supplemented with 1.8×10^{-4} M adenine, 10% fetal calf serum (FCS), $5\,\mu\mathrm{g\,ml}^{-1}$ insulin, $0.4\,\mu\mathrm{g\,ml}^{-1}$ hydrocortisone, 8.4 $\mathrm{ng\,ml}^{-1}$ cholera toxin, and $10\,\mathrm{ng\,ml}^{-1}$ EGF at $37\,^{\circ}$C for 48 h.

Routine culture of cell lines

HT29 and 3T3-K fibroblasts are maintained in DMEM with 10% FCS while SVK14 cells were grown in RPMI 1640 and 10% FCS. B9 cells (kindly provided by L. Aarden, University of Amsterdam, The Netherlands) are maintained in RPMI 1640 supplemented with 10% FCS and 20 units/ml of recombinant IL-6 (Amersham).

Production of collagen-coated plates

Type 1 (rat-tail) collagen is prepared using the method of Bell, Ivarsson and Merrill (1979) and stored at $4\,^{\circ}$C prior to use (see also method of Fusenig, Chapter 10). The collagen solution is applied to the surface of tissue culture plates, then removed by aspiration; the plates are then allowed to dry at room temperature overnight under a UV light to sterilize them.

IL-6 assay

Prior to assay, B9 cells are washed twice and resuspended in RPMI 1640 and 10% IL-6 free FCS. Cell number is determined and viability is checked by trypan blue exclusion. For each assay the cell viability must be greater than 95%. The cells are plated out at a density of 2×10^6 per $25 \, \text{cm}^2$ tissue culture flask (Falcon) and incubated overnight at $37 \, °C$ in 5% CO_2/air mixture.

Cells to be tested for IL-6 production are incubated in RM + for 72 h; the medium is then centrifuged at 1800 g for 5 min to remove any cells and can be stored at $-20 \, °C$ if necessary. After removal of the medium for testing, cells are washed twice with 0.02% EDTA in calcium and magnesium-free phosphate buffered saline at pH 7.0 (PBSA) and then dissociated at $37 \, °C$ with 0.25% trypsin diluted 1:4 in 0.02% EDTA for 10–20 min. Cell numbers and viability are determined (viability in each case should be greater than 95%).

The assay for IL-6 is carried out in 96 well flat-bottomed (Costar) microtiter plates and involves, in triplicate, serial 2-fold dilutions of the test supernatants in RPMI 1640 supplemented with 10% FCS, the final volume in each well being 100 μl. 10^4 B9 cells in RPMI 1640 supplemented with 10% FCS are added to each well bringing the total volume to 200/μl. The plates are then incubated at $37 \, °C$ in a humidified 5% CO_2/air mixture. After 72 h, proliferation of the B9 cells is measured by MTT assay.

MTT (Sigma) is dissolved in PBSA at 5 mg/ml and filtered for sterility and to remove the small amount of insoluble residue that is present in some batches. 10 μl of MTT is added to each well of the assay plates and the plates are incubated at $37 \, °C$ for 4 h. 100 μl of 10% SDS, dissolved in 0.01N HCL, is added per well and incubated overnight at $37 \, °C$ in a humidified 5% CO_2/air mixture, to dissolve the blue formazan crystals. The plates are read on a Titertrek Multiskan Plus MKII microelisa plate reader at a test wavelength of 570 nm. Results are expressed in units/ml where 1 unit/ml is the concentration of recombinant IL-6 that leads to half maximal absorption at that wavelength.

Part III

METHODS FOR STUDYING DIFFERENT SUBPOPULATIONS OF KERATINOCYTES

Introduction

Both intact epidermis and keratinocyte cultures consist of a mixture of dividing cells, and cells at different stages in the terminal differentiation pathway. This section describes a number of methods for comparing different subpopulations of keratinocytes. *In situ* hybridization (see Fisher, Chapter 21) and electron microscopy (see Eady *et al.*, Chapter 22) can be used to compare keratinocytes without disrupting the spatial organization of the cells. Suspension in methyl cellulose (see Watt, Chapter 20) is a simple technique for inducing premature terminal differentiation of keratinocytes. The other protocols, elutriation (see Teumer *et al.*, Chapter 17), unit gravity sedimentation (see Rizk-Rabin and Pavlovitch, Chapter 18) and flow cytometry (see Jones and Watt, Chapter 19), involve physical separation of dividing and differentiating cells and exploit the changes in keratinocyte size and buoyant density that occur as part of the differentiation process.

17   Isolation of RNA from keratinocytes of different sizes fractionated by elutriation

J. TEUMER, K. ZEZULAK and H. GREEN

As keratinocytes undergo terminal differentiation, they enlarge (Sun and Green 1976; Watt & Green, 1981) and amongst basal cells differences in size correlate with proliferative capacity (Barrandon & Green, 1985). The method that we describe allows the fractionation of different size classes of keratinocytes in sufficient quantities for RNA isolation and Northern blotting. Results obtained using the method are described in the *Keratinocyte Handbook* (Teumer *et al.*, 1994). The technique has been used by others to study gene expression in keratinocytes (Di Marco *et al.*, 1991; Gherzi *et al.*, 1992) and has been used to separate cell populations enriched in melanocytes (D'Anna *et al.*, 1988).

Cell culture

Human foreskin keratinocytes (passage 3–4) are cultured with supporting irradiated 3T3 cells according to published procedures (Rheinwald & Green, 1975; Simon & Green, 1985). For elutriation, cells are plated at $(4–5) \times 10^5$ cells per 100 mm dish and harvested 7 days later, when they are 1–2 days postconfluent. At this time, most of the 3T3 feeders have detached from the stratified cell layer.

Keratinocyte fractionation

Keratinocytes are harvested by trypsinization, centrifuged and resuspended in medium containing DNase I at 40 ng/ml to reduce cell aggregation. They are then collected by centrifugation and resuspended in 30 ml of DME medium containing 5% bovine serum. Fractionation is performed using a Beckman JE-6B elutriation system. After assembly of the rotor and elutriation apparatus, medium is pumped through the system with the rotor spinning at 1000 rpm at room temperature, and air bubbles are purged. The cell suspension is loaded into the elutriator at 6.5 ml/min, a flow rate that allows cells to fill the elutriation chamber to its widest part (as monitored through the viewing port on the centrifuge door). During the early stages of loading, many cells of all sizes wash through the chamber and into the collection tube. When the elutriate becomes clear of cells, the flow rate is increased in a stepwise manner by increments of 2.5 ml/min for the first six fractions, then by increments of 5 ml/min for the remaining fractions. At each step, the progress of the elutriation is monitored and when the elutriate becomes clear of cells, the flow rate is increased. Usually 12–15 fractions of volume 100–200 ml are collected until the largest cells are elutriated. The number of cells in each fraction is counted and the cells are collected by centrifugation and resuspended in a volume that gives a cell concentration such that, when the suspension is introduced into a hemacytometer, 100–200 cells appear in a microscope field. The cells are photographed under phase contrast at 10× magnification. The pictures obtained in this manner are examined with a measuring magnifier to estimate cell diameters. Some adjacent fractions are pooled to give a total of six fractions. In the first fraction 97% of the cells are 11–14 μm in diameter and the majority are clonogenic, while in the sixth fraction over 91% of cells have a diameter greater than 20 μm, are unable to found colonies and may possess cornified envelopes (see Teumer *et al.*, 1994).

RNA extraction

Each fraction is centrifuged, and cell pellets are dissolved in 3.5 ml of 5M guanidinium isothiocyanate. The viscous solution is passed through a 21-gauge needle 4–5 times to shear the DNA, and is layered over 1.5 ml of 5M CsCl in SW50.1 polyallomer centrifuge tubes. The samples are centrifuged at 35K rpm for 12–18 h at room temperature (Chirgwin *et al.*, 1979). The pelleted RNA is recovered and measured by spectrophotometry. Poly[A]$^+$ RNA is selected by two rounds of oligo[dT]-cellulose chromatography (Sambrook *et al.*, 1989).

Northern blots

Northern blots of poly[A]$^+$ RNA are performed according to published procedures (Sambrook *et al.*, 1989). Between 0.5 μg and 2 μg of poly[A]$^+$ RNA per fraction are separated on formaldehyde–agarose gels. DNA probes are labelled with ^{32}P using the random priming method (Feinberg & Vogelstein, 1983). The radioactive signals on the blots are visualized and quantitated using a Molecular Dynamics PhosphorImager.

18   Unit gravity sedimentation technique

M. RIZK-RABIN and J. PAVLOVITCH

Cells, nuclei, and chromosomes sediment at ambient gravity at velocities varying from a few centimeters per hour to fractions of a millimeter per hour. Assuming that the 'streaming limit' is not exceeded, rapid separation of keratinocytes at unit gravity rests primarily on one technical problem: the undisturbed layering of a thin film of particle suspension on an otherwise undisturbed liquid column. Such a thin film guarantees a spatial separation of keratinocytes that differ slightly in volume, within a reasonably short time. The use of unit gravity to separate subpopulations of basal keratinocytes and to separate basal from differentiated cells has been reported (e.g. Pavlovitch *et al.*, 1991; Nicholson & Watt, 1991). The procedure that follows is used to separate different subpopulations of keratinocytes isolated directly from newborn rat skin.

Preparation of epidermal cells

(i)   The skin is removed from each animal, stretched and floated on 0.25% trypsin, at 4 °C for 15 h.
(ii) The epidermis is separated mechanically from the dermis using forceps. Gentle pressure is applied to the dermis to remove the lowermost layer of epidermal cells.
(iii) Single cell suspensions are prepared by shaking the epidermis in 10% decalcified fetal calf serum (FCS) in calcium-free modified Eagle's medium (MEM) at 4 °C for 1 h. 0.25% deoxyribonuclease 1 (Sigma) is added to the solution to avoid aggregation.
(iv) Cell suspensions are filtered through three layers of gauze then through 200 um nylon mesh, sedimented through a layer of 10% FCS in MEM, and resuspended in 0.5% Ficoll in PBS.

The sedimentation chamber

The sedimentation chamber consists of two parts made of Perspex: a cylindrical part, the bottom of which has a shallow cone and a rotational symmetric rhomb fixed in the cone, and a top cone which can be removed for cleaning purposes with again a rotational symmetric rhomb fixed in the cone. As these rhombs act as flow deflectors, liquid entering or leaving the sedimentation chamber from (u) or (d) flows symmetrically through the narrow space between the deflectors and the cones. Thus, the high

Keratinocyte methods by Irene Leigh and Fiona Watt
© Cambridge University Press, 1994, pp. 101–103

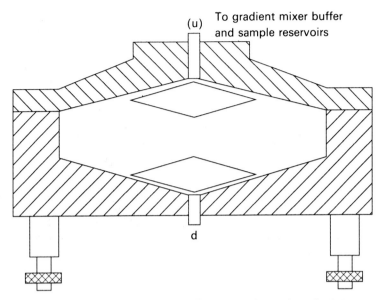

Fig. 18.1.   Unit gravity Sedimentation Chamber. (Order from P. K. Lens, Gesel Instrument Maker, Schutzlius 34, 1186 2 Amsterdam, Holland, Tel. 20410250.)

initial velocity of flow is lost and undisturbed layering of liquids in the chamber is permitted. This reduces turbulence and minimizes the distortion of gradients and cell fractions.

The chamber cannot be autoclaved. For sterilization, 6% Deconex solution or equivalent is used, followed by rinsing with sterilized water. To avoid attachment of cells to the plastic walls of the chamber, we rinse the chamber with a 1% Triton × 100 solution just before the last rinse with sterilized distilled water.

Before filling the chamber, make sure it stands horizontally, by using adjustment screws and a spirit level.

Mode of operation

(i)   The closed chamber is filled completely via (d) with dense 'cushion' liquid (fluorinated hydrocarbon, Jansen France) at a rate of 15 ml/min. The material of the cushion liquid should be identical with that of the density gradient (or should at least have the same diffusion constant).

(ii)   360 ml gradient of Ficoll 70 (Pharmacia) (8–2%) in PBS is introduced into the chamber via (u) over the course of 30 min.

(iii)   A 30 ml density gradient of Ficoll (1–2%) in PBS (buffered step gradient, known to stabilize sample layers against the streaming phenomenon) is introduced into the chamber via (u) by letting out cushion liquid via (d). This takes 5 min.

(iv) 20 ml of sample is introduced into the chamber via (u), followed by an overlay of PBS buffer (25 ml); this step takes 20 s. While this operation is taking place, cushion liquid is let out via (d).

(v) Within the following 2 min, the process of letting out cushion liquid via (d) is continued until the film of particles has reached the cylindrical part of the sedimentation chamber (surface area 177 cm^2). An extremely thin band of cells (0.5 mm) is obtained.

(vi) Now there is an extremely thin film of undisturbed sample material over the density gradient, the cells are allowed to sediment at unit gravity. We usually leave them for 2 h.

(vii) At the end of the procedure, cushion liquid is introduced via (d) and the gradient liquid is collected from the top of the chamber. We collect 16 fractions of 22.5 ml.

Note

(i) The times and the volumes mentioned above should be considered as approximative values and might vary according to the populations to be separated.

(ii) All the processes of introducing or removing solutions can be controlled simply by varying the level of the cushion fluid.

19 Applications of flow cytometry in the study of keratinocytes

P. H. JONES and F. M. WATT

Flow cytometry allows the rapid quantitative measurement of the fluorescence of labelled cells in suspension. Cells are passed one by one through a laser beam. Detectors then measure the amount of scattered light in the forward direction (FSC), light scattered to the side (SSC) and the fluorescence of the cell. The data are stored on computer. Up to three fluorescent labels of different colors can be measured simultaneously in the same cell. Commonly used fluorochromes are fluorescein, which emits green light, R-Phycoerythrin, emitting orange light, and 'Tricolor', which emits red light. Over 1000 cells can be analysed per second (Ormerod, 1990).

Applications of flow cytometry in the study of keratinocytes

Differentiation

Basal and suprabasal cells can be identified and the level of expression of specific antigens in each population measured (Watt & Jones, 1992).

Cell cycle studies

DNA content can be measured with propidium iodide, RNA content with acridine orange and the proportion of actively cycling cells in a population determined using bromodeoxyuridine (Staiano-Coico et al., 1986; Pavlovitch et al., 1989; Khandke et al., 1991; Ormerod, 1990).

Other cell characteristics

Fluorescent probes can be used to measure parameters such as membrane fluidity, intracellular calcium levels, pH, and enzyme activity (Hachisuka, et al., 1990; Ormerod, 1990).

Cell sorting

Cells can be recovered from a flow cytometric cell sorter (FACS) under sterile conditions and plated in culture to measure colony forming efficiency (Jones & Watt, 1993); cells can also be extracted for isolation of protein, RNA and DNA.

Limitations of flow cytometry

A single cell suspension is required.

Harvesting keratinocytes requires the use of trypsin and therefore trypsin–sensitive epitopes on the cell surface are lost. Although cells can be identified as basal or suprabasal on the basis of FSC and SSC or using trypsin–insensitive markers (Watt & Jones, 1992; Jones & Watt, 1993), it is not possible to say which suprabasal layer of the culture the cells originally came from as trypsinization destroys the spatial orientation of the cells.

Loss of information on the pattern and location of staining.

The distribution of staining within a cell is not measured by the flow cytometer. Nuclear, cytoplasmic and membrane staining will all appear identical to the machine.

Cell size

Much of the literature on flow cytometry deals with cells which are uniform in size. Keratinocytes vary in diameter from 8 μm to over 25 μm (Watt & Green, 1981). A flow cytometer measures the total fluorescence of each cell independent of its size. Under the microscope, a large cell with a low number of fluorescent molecules bound per unit area would look much duller than a small cell which bound the same number of fluorochrome molecules. The flow cytometer would record both these cells as having the same fluorescence. This can make the interpretation of a histogram of the whole keratinocyte population difficult and emphasizes the need to use the computer to analyze the small basal cells and the much larger suprabasal cells separately.

Cell viability

Sorting cells for a period of longer than 1 hour presents a problem as the viability of cells falls sharply after this time. This sets a limit on the number of cells that can be sorted. Numbers of sorted cells in excess of 5×10^6 are difficult to obtain.

Protocols

The following points are important in preparing keratinocytes for flow cytometric analysis.

Cell clumping

Keratinocytes in suspension readily form clumps which can block the machine. Cells should be kept at 4 °C and filtered through nylon mesh

(60–100 μm pore size). Cultures should be preconfluent or have just reached confluence when harvested, as it is much more difficult to make a single cell suspension from a postconfluent culture.

Removing feeder cells

When harvesting keratinocytes, it is essential to remove the 3T3 feeder cells first, as 3T3 cells have the same light scattering characteristics as suprabasal keratinocytes.

Viability

The shorter the time taken in preparing the cells, the better the viability. Adequate antibody binding can usually be achieved in a 10-minute incubation. Using antibodies directly conjugated to a fluorochrome saves time and enables two- and 3-color experiments to be performed. The procedure for conjugation of antibody to fluorescein isothiocyanate (FITC) is very straightforward (Goding, 1976).

Using a flow cytometer

If unfamiliar with the technique, obtain the assistance of an experienced operator in setting up the machine.

Double labelling of keratinocytes with peanut lectin (PNA) and antibodies to integrins

PNA binds selectively to suprabasal cells and anti-integrin antibodies bind selectively to basal cells (Watt & Jones, 1992).

Harvest keratinocytes with trypsin/versene and wash in serum-containing medium.

Resuspend cells in PBS containing 1 mM $CaCl_2$ and 1 mM $MgCl_2$(PBSABC) at 4 °C, pipetting several times to disaggregate cell clumps. Filer through a 63 μm nylon mesh (R. Cadish and Sons, London N3, UK).

Place 10^5 cells in a volume of 100 μl in 5 ml tubes. Prepare the samples as shown in Table 19.1.

Saturating concentrations of each reagent should be determined by titration

Incubate for 10 min on ice with a saturating concentration of first layer antibody and/or lectin. Wash the cells once in 5–15 ml cold **PBSABC** and incubate with second layer reagents as shown in the table.

Wash the samples once more and leave in a final volume of 100–200 μl. If the FACS is available samples can be analyzed immediately. Propidium

Table 19.1. Samples needed in two color labelling for one integrin subunit and PNA

	First layer			Second layer	
Sample	CD 29	CD 3	Biotinylated PNA	SAM FITC	Tricolor
Test	+		+	+	+
Control 1	+			+	
Control 2			+		+
Control 3		+	+	+	+
Control 4	+			+	+
Unstained					

Note: Reagents used–

- Anti-CD29 (Jansen) is a mouse anti-$\beta 1$ integrin antibody.
- Anti-CD3 (Jansen) is the same species and isotype as anti-CD29 but does not bind to keratinocytes and is used as a negative control.
- Biotinylated PNA (Vector).
- SAM-FITC - Sheep anti-mouse IgG, FITC conjugate (Sigma).
- Tricolor is used as a streptavidin conjugate (Caltag).

iodide (PI) can be added just before analysis to a final concentration of 5 ug/ml; this causes nonviable cells to fluoresce an intense red, enabling them to be identified and excluded from analysis. If the machine is not available immediately, cells may be fixed in 1% paraformaldehyde (PFA) and kept at 4 °C in the dark. PI should not be added to fixed cells as they will all be positive.

The FACS is set up using all the samples, and information should be acquired on 10 000 cells from each of the stained samples. Tricolor emits in the far red. With appropriate filters in the cytometer, the signal from Tricolor can be separated from the FITC signal. Typical results are shown in Fig. 19.1.

Use of light scattering characteristics to separate basal and suprabasal cells

Basal and suprabasal cells may also be identified by their light scattering characteristics (Fig. 19.2). Fixation alters the pattern of light scattering although if 1% paraformaldehyde is used the changes are minimal.

Comparing subpopulations of keratinocytes

Because of the variability in cell size in the keratinocyte population it is advisable to sort cells from different subpopulations and measure their mean diameter. Dividing the mean fluorescence by the square of the mean diameter gives an estimate of the relative number of fluorochrome molecules

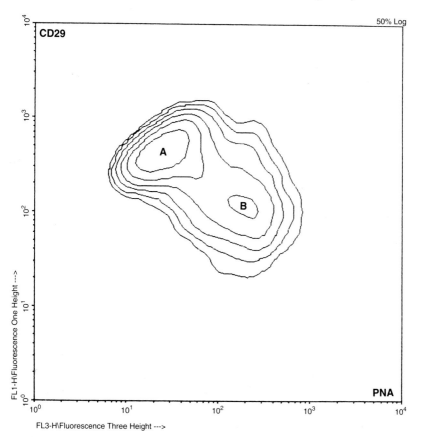

Fig. 19.1. Contour plot of cells stained for the $\beta 1$ integrin subunit and PNA. A contour plot is a two-dimensional histogram. The fluorescence due to PNA is shown on the x-axis and that due to $\beta 1$ integrin on the y-axis, both in arbitrary units plotted on a log scale. The number of cells with a given fluorescence value is indicated by the contours: the more contour lines the greater the number of cells. There are two populations of cells. Population A contains cells dull in PNA fluorescence (basal cells) which stain brightly for $\beta 1$ integrin. Population B comprises cells which are bright in PNA (and therefore suprabasal) but are dull in $\beta 1$ fluorescence.

bound per unit area of the cell membrane, which is more likely to be of biological significance than the total number of fluorochrome molecules bound.

Sterile sorting

It is vital to sterilize the machine for at least 1 hour with detergent (e.g. a 250-fold dilution of 7 X, Flow Laboratories) which should then be washed

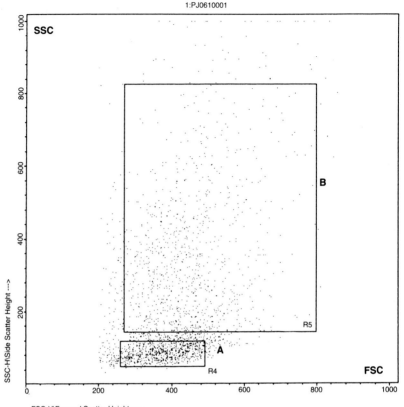

Fig. 19.2. Dot plot showing the light scatter of cultured keratinocytes in a forward direction (FSC) on the *x*-axis and light scattered to the side (SSC) on the *y*-axis, both measured in arbitrary units. Each dot represents one cell. When sorted, cells in region A are shown to be basal cells: they are less than 13 microns in diameter, do not express involucrin and are able to form colonies. Region B contains suprabasal cells: the cells are larger, over 50% are involucrin positive and they have a very low colony forming efficiency (Jones & Watt, 1993).

out and followed by 70% ethanol for 1 h. Finally an antibiotic solution (e.g. gentamicin 5 mg/ml, 2500 international units/ml penicillin and 2.5 mg/ml streptomycin, Gibco) should be run through for 30 min. Antibodies and PI can be sterilized with a 0.22 μm filter (Costar).

5×10^6 cells are needed for a typical sterile sort, e.g. to determine the colony forming ability of keratinocyte subpopulations (see Jones & Watt, 1993). Samples should be filtered twice with autoclaved Nylon mesh, once before staining and once immediately prior to sorting. Cell viability should be at least 90%. The possibility of contamination is minimized and colony forming efficiencies improved if keratinocytes are sorted directly into medium in culture dishes pre-seeded with feeder cells. If a Becton–Dickinson FACStar plus flow cytometer is used, this is easily done with the Automatic Cell Deposition Unit.

20   Suspension-induced terminal differentiation of keratinocytes

F. M. WATT

Introduction

Keratinocytes can be induced to undergo terminal differentiation by placing them as a single cell suspension in medium made viscous by the addition of methyl cellulose (Green, 1977; Watt, Jordan & O'Neill, 1988; Adams & Watt, 1989). Extracellular matrix proteins and antibodies to integrin subunits can be added to the methyl cellulose in order to inhibit differentiation (Adams & Watt, 1989; Watt et al., 1993).

Preparation of methyl cellulose

(i)   Add 3.5 g methyl cellulose powder (Aldrich Cat. No. 27,441-0) and a magnetic stirring bar to a flat-bottomed plastic centrifuge bottle with a screw cap. Autoclave.

(ii) Add 180 ml serum-free medium, preheated to 60 °C. Stir at room temperature for 30 min.

(iii) Transfer to 4 °C and stir overnight. Solution will now be clear and viscous.

(iv) Add 20 ml serum (for 10% final concentration) and any other medium supplements.

(v)   Centrifuge at 9500 rpm in a Beckman J2-21 at 4 °C for 30 min.

(vi) Decant supernatant into sterile bottles and store at 4 °C (or at − 20 °C if not to be used within a week).

Addition of keratinocytes

A single cell suspension of keratinocytes is prepared in complete culture medium at a concentration of 10^6/ml. Cells are added to the methyl cellulose medium to a final concentration of 10^5/ml, mixed thoroughly and decanted into bacteriological plastic petri dishes coated with 0.4% polyHEMA (Watt et al., 1988) to prevent cell attachment. To harvest cells from suspension, the methyl cellulose is diluted 10-fold with PBS and the cells are recovered by centrifugation at 1000–2000 rpm for 5 min.

Note

The methyl cellulose medium is extremely viscous and is most easily dispensed by pouring or using a positive displacement pipettor. Cells can be mixed into methyl cellulose medium by stirring with a sterile pipette tip.

Keratinocyte methods by Irene Leigh and Fiona Watt   113
© Cambridge University Press, 1994, p. 113

21   *In situ* hybridization

C. FISHER

Introduction

In situ hybridization can be used to localize specific mRNAs in cultured keratinocytes and in histological sections of skin. The technique that is described below gives good results with sections of mouse skin. The differences between different *in situ* hybridization methods usually lie in section preparation and probe detection. My protocol uses paraffin sections, but it is also possible to use frozen sections; the tissue can be fixed before freezing, or fixed after sectioning. I detect probes with ^{35}S-UTP; alternatives are to use a [3H] label or nonisotopic labels such as digoxigenin. For a range of protocols see Wilkinson (1992).

Materials and methods

Fixation

Fix tissue in 4% paraformaldehyde on ice for 1–5 h, dehydrate in an ethanol series, and embed in paraffin. Fixation and subsequent steps may vary with tissue size. Alternatively, unfixed tissue can be snap frozen. Frozen sections should be air dried at room temperature, fixed for 1–5 min in 4% paraformaldehyde, and stored in 70% ethanol at 4 °C until use.

RNase precautions

Caution should be exercised throughout these procedures to avoid RNase contamination. Gloves should be worn at all times to avoid contamination by finger RNases; only autoclaved glassware (or sterile, disposable plastic ware) should be used; and all solutions should be autoclaved (except for the alcohols). DEPC treated water may be used but we have obtained good results with autoclaved, deionized water as well.

DEPC treating water

Add 1 ml diethyl polycarbonate (DEPC; Sigma) per liter H_2O with vigorous stirring until it goes into solution (10–20 minutes), autoclave, and then continue stirring as water cools to allow ethanol to blow off.

Silane coating slides

We have obtained particularly consistent and good results, and lost very few sections, with the use of 3-aminopropylsilane coated slides to support our sections (Rentrop *et al.*, 1986). Slides should be acid cleaned, rinsed and dried, immersed for 5 s in a freshly prepared 2% solution of 3-aminopropylsilane in dry acetone, washed briefly 2 × in dry acetone, rinsed in diH$_2$O, and dried overnight at 4 °C. These slides can be used for either paraffin or frozen sections and can be stored indefinitely at room temperature.

A note of caution: avoid pseudofrosted slides. Some slides have been treated with a plastic resin to resemble true frosted slides and this resin does not stand up to some of the rinse steps following hybridization. Remember: wear gloves!.

Sectioning and mounting

Paraffin sections of 5–10 μm should be mounted on silane coated slides by floating on autoclaved water on a slide warmer. Make sure that section is positioned toward the end of the slide (but not close to the edge) so it will be easy to cover with emulsion during the autoradiography step (see below). Let slides dry for 1–2 h and remove to a slide box. These can be kept for months at room temperature. Frozen sections should be mounted and stored frozen, or fixed in paraformaldehyde and stored in 70% ethanol at 4 °C (see above).

Riboprobe synthesis

The following solutions should be prepared in sterile water, aliquoted, and frozen.

5 × transcription buffer
200 mM Tris-HCl, pH 7.5
30 mM MgCl$_2$
10 mM spermidine
100 mM NaCl

5 × rNTPs
ATP, GTP, CTP, 2.5 mM each (omit labelled nucleotide)

Also have on hand
0.2 M DTT (make up fresh)
RNasin or placental ribonuclease inhibitor
^{35}S-UTP (12.5 mCi/ml)
Appropriate RNA polymerases (SP6, T7, T3)

The reaction mix should be assembled in a microfuge tube in which ^{35}S-UTP has been lyophilized. Divide 1 mCi 35S-UTP (in 80 μl) into 4 separate 1.5 ml autoclaved tubes (20 μl, or 250 μCi/tube) and lyophilize; this generally results in an $\sim$ 18 μM concentration for the limiting nucleotide. In these tubes, combine the following at room temperature (not on ice or spermidine may precipitate with nucleic acids) for a typical reaction:

5 × transcription buffer	2.0 μl
0.2 M DTT	0.5 μl
RNasin or placental ribonuclease inhibitor	0.5 μl
(0.5–2 units/ll final concentration)	
^{35}S-UTP (> 10–12 M)	dried in tube
5 × rNTPs	2.0 μl
DNA (linearized, 1 μg/μl	1.0 μl
RNA polymerase (SP6, T7, T3; 5–10 U)	1–2 μl
	7–8 μl subtotal
H$_2$O (DEPC treated)	2–3 μl
Total	10.0 μl

Incubate at 37 °C for 1–2 h.

Removal of DNA template
(Optional–some say this step is not necessary as DNA will not denature under hybridization conditions).

Add 10 units RNasin or placental ribonuclease inhibitor and 0.3 μg DNase (RNase-free). Incubate at 37 °C for 10 min.

Purification of RNA transcripts
(i)   Add to the reaction mixture:
  • 4 μl t-RNA (10 mg/ml; Proteinase K treated and chloroform extracted)
  • 56.0 μl DEPC-treated water
  • 50.0 μl 2% SDS
  • 100.0 μl 0.6 M sodium acetate (pH 5.2).
(ii) Vortex briefly, add 100 μl phenol, vortex, and add 100 μl chloroform. Vortex again and separate phases by centrifugation (1 min in microfuge). Remove upper aqueous layer to fresh, sterile tube and reextract with 200 μl chloroform. Again, separate phases by centrifugation and remove aqueous layer to fresh tube being careful to leave behind the flocculent interface. Precipitate RNA by adding 750 μl cold ethanol and chill in dry ice/ethanol slurry for 5 min. Microfuge for 5 min and briefly drain pellet. Resuspend pellet in 200 μl of 0.2% SDS, 2 mM EDTA, and 0.3 M ammonium acetate (pH 5.2). Add 2 volumes ice

cold ethanol and precipitate as above. Collect pellet and repeat resuspension/precipitation 2 ×.

(iii) Dry final pellet and resuspend in sterile, DEPC-treated water. We usually get 0.5 to 2.0 × 10^8 cpm of RNA probe.

The optimal probe size is 150–200 bp. The size of one's probe can be easily reduced by hydrolysis at pH 10.2. The reader is referred to Angerer et al. (1987). for details.

Pretreatment of sections and hybridization

Have the following buffers prepared and autoclaved:
- 100 mM Tris, 50 mM EDTA, pH 8.0
- Tris, 100 mM glycine, pH 7.0
- 100 mM triethanolamine, pH 8.0 with HCl
- 2 × SSC (0.3 M NaCl, 0.03 M Na citrate, pH 7.0), 3-4 litres
- 25 × SSC
- 10 × Denhardt's (0.2% Ficoll, 0.2% polyvinyl pyrrolidone, 0.2% BSA)

Deparaffinization (if necessary) and rehydration
Use routine histological procedures (two changes xylene to deparaffinize and graded EtOH series for rehydration) except use autoclaved glassware and fresh solutions. After 70% EtOH, soak slides in two changes of 2 × SSC for 2 min.

Protease treatment
Incubate slides for 5 min in 1 μg ml solution of proteinase K in Tris-EDTA, pH 8.0 buffer. Rinse in 2 × SSC briefly. (Note: the time used varies with tissue; several time points should be used with initial attempts. I have eliminated this step from my normal procedure as comparable or better results may be obtained without it.)

Acetylation and glycine treatments
Soak slides in 225 ml of triethanolamine buffer containing 562.5 μl acetic anhydride (be careful with this, it can cause burns). Mix well before adding slides. Rinse twice in 2 × SSC. Place slides in Tris-glycine buffer for 30 min at room temperature. Rinse in 2 × SSC and hold for hybridization.

Prehybridization
Prepare fresh solution of 0.5 M DTT. Mix 80 μl 25 × SSC and 20 μl DTT solution. Boil DNA and t-RNA for 2 min just prior to preparation

of mixture. For 1 reaction prepare the following:

Formamide	5 μl
20 × SSC-100 mM DTT	1 μl
calf thymus t-RNA (10 mg/ml)	1 μl
salmon sperm DNA (10 mg/ml)	1 μl
10 × Denhardt's	1 μl
sterile H$_2$O	1 μl
Total	10 μl

Apply 10–20 μl to each section, coverslip with autoclaved coverslips, place in humidified chamber, and incubate at 45 °C for 1 h.

Hybridization
Prepare hybridization mixture identical to prehybridization mixture but substitute 1 μl of boiled ^{35}S-labelled RNA probe ($\sim 1 \times 10^6$ cpm/μl) for water. Carefully remove coverslip from section, blot away excess pre-hybridization mixture, and apply 10–15 μl of hybridization mixture to section. Coverslip with sterile coverslip, seal with Carter's rubber cement, place in a humidified chamber, and hybridize for 3 h to overnight at 45 °C.

Formamide rinses and RNase treatment

Carefully peel away rubber cement and remove coverslips by gentle agitation in 2 × SSC in a 50 ml Falcon tube. Hold in 2 × SSC until all slides are finished. Transfer to 2 × SSC–50% formamide at 52 °C for 25 min. Agitate occasionally. Wash 4 times in 2 × SSC. Meanwhile, prepare RNase solution in 2 × SSC: 100 μg/ml RNase A and 1 μg/ml RNase T. (RNase should be purchased in liquid form, if possible, to avoid introducing the lyophilized enzymes to the lab.) Blot excess fluid from around sections, apply RNase solution, coverslip, and incubate in humidified chamber for 30 min at 37 °C. Remove coverslips by agitation in plastic tube containing 2 × SSC. Avoid introducing RNase to the lab. Always use same glassware, preferably the same as formamide glassware. Discard plastic tube. Transfer slides to 50% formamide-2 × SSC for 5 min agitating occasionally.

Rinse in 2 × SSC, then rinse on shaker in 2 × SSC for 10 min at room temperature. Dehydrate through graded ethanol series to 95% ethanol and dry under a gentle stream of air. The slides are now ready for auto-radiography.

Alternative rinses and RNase treatments (particulary for use with thiolated probes)

After removing rubber cement agitate slides in 4 × SSC, 10 mM DTT in a 50 ml Falcon tube until coverslip falls off. Incubate in 4 × SSC, 10 mM DTT for at least 1 h.

Wash in 0.1 × SSC, 10 mM β-mercaptoethanol for 15 min at 50 °C. (This high stringency wash approaches the stringency of the hybridization conditions.)

Wash in 0.1 × SSC for 30 min at room temperature.

Autoradiography

As long as a few simple procedures are followed, there is no reason to be anxious about autoradiography. The problems you encounter with background will more than likely be due to nonspecific radioactivity rather than darkroom technique. It helps to keep clean glassware for use with autoradiography separate from other glassware. The NTB-2 emulsion is insensitive to standard darkroom lighting such as a safelight equipped with a Wratten red No.1 filter. However, whenever possible, we do procedures, such as drying, in absolute darkness.

When emulsion is received melt it in a 4 °C water bath, mix it 1:1 with 0.6 M ammonium acetate, and store in 10–20 ml aliquots at 4 °C. Make sure these are stored in an easy to retrieve manner (it helps to individually wrap the samples) under light-tight conditions. When ready to dip slides, remove emulsion aliquot from refrigerator and warm to 45 °C for 30–40 min in a water bath set up in the darkroom. Carefully pour emulsion into dipping chamber and slowly and evenly immerse slide 2–3 × into the emulsion. (Dipping chambers are available through microscopy supply houses or one can easily be made by sawing off a 100 ml graduated cylinder at the 30 ml mark.) Blot the dipped slides on paper towels and allow to dry for 45 min while standing on their ends in a suitable rack. Transfer the dried slides to light-tight slide boxes along with some desiccant wrapped in tissue paper. Put the slide boxes into another container (used film or photographic paper boxes are useful) and store at 4 °C for 1–5 days.

Development

Remove box containing slides from the refrigerator and allow to come to room temperature ($\sim$ 30–45 min). Transfer the slides to a glass slide holder in the darkroom and develop for 2 min in D-19 developer. Stop the reaction with 1% acetic acid for 30 s and fix in Kodak Rapid Fixer for 5 min. From here on, the slides can be handled in the light. We generally do not counterstain when first testing probes because it can interfere with viewing a faint signal. However, sections can be countersigned by a number of methods; we prefer to briefly stain with hematoxylin (5 s). After rinsing well, dehydrate the slides in a graded series to 100% ethanol, clear in xylene (2 ×, 3 min each), and coverslip with Permount.

22   Electron microscopy of keratinocytes

R. A. J. EADY, A. ISHIDA-YAMAMOTO,
and J. A. McGRATH

Electron microscopy

The methods used for processing skin or keratinocytes, either as cultured colonies or as isolated cells, are generally well established. Details of our preferred techniques have been described previously (Eady, 1985). Briefly, most skin samples will be fixed by immersion rather than by perfusion, and the time taken between removing the sample from the patient or experimental animal and immersing it in fixative should be as short as possible. We recommend sequential fixation involving primary fixation in a buffered aldehyde solution such as half-strength Karnovsky's fixative containing a mixture of formaldehyde, which is prepared from para-formaldehyde, and glutaraldehyde, in cacodylate buffer. Glutaraldehyde has strong cross-linking properties but penetrates the tissue poorly. Form-aldehyde penetrates faster but is less effective in preserving fine structure.

Secondary fixation in a 1–2% solution of osmium tetroxide in distilled water is recommended but not always necessary. If the same blocks are to be used for immunoelectron microscopy (see below), it would be preferable to omit the osmium.

Because glutaraldehyde and osmium penetrate tissue poorly, the specimens should be cut up into small pieces (<2 mm^3) or thin slivers not more than 1 mm thick. The specimens are then dehydrated through graded concentrations of ethanol and embedded in resin – usually Epon or Araldite.

The method chosen for processing cell suspensions or cell cultures will depend to some degree on the experimental requirements. If it is desirable to examine the cultures without disturbing their attachment to the dish or coverslip, special care has to be taken to process the cells *in situ* (see Eady, 1985).

Immunoelectron microscopy

The three main strategies for preparing specimens for immunoelectron microscopy include postembedding and pre-embedding methods, and thirdly, the use of cryoultrathin sections. Each method has particular strengths and weaknesses which are discussed in more comprehensive reviews (see, for example the texts edited by Bullock & Petrusz, 1982; and by Polak & Priestley, 1992).

Keratinocyte methods by Irene Leigh and Fiona Watt
© Cambridge University Press, 1994, pp. 121–124

Postembedding method

This is the recommended procedure for labelling intracellular antigens, including keratins, profilaggrin/filaggrin, involucrin, and loricrin. The full procedure including incubations and washes are carried out using ultrathin sections after they have been collected on grids. Epoxy-resin embedded sections are often unsuitable because the antigens are masked or destroyed during processing. However, certain antigens can be labelled after the osmium has been removed using an oxidizing agent such as sodium metoperiodate (Bendayan & Zollinger, 1983; Merighi, 1992).

In planned experiments, it is preferable, if not essential, to embed the specimens in a water-soluble resin such as LR White (Newman & Hobot, 1987) or Lowicryl K4m or K11m (Carlemalm, Garavito & Villiger, 1982) that has been developed specifically for immunocytochemistry. The Lowicryl resins are polymerized at low temperature by ultraviolet radiation. For most purposes the specimens can be lightly fixed in an aldehyde fixative, usually formaldehyde (from paraformaldehyde) in concentrations up to 4%, but with very low concentrations of glutaraldehyde (not more than 0.5%). Alternatively, periodate-lysine-(para)formaldehyde fixative (McLean & Nakane, 1974) is especially good for fixing carbohydrates and lipids.

We have adopted a protocol (Shimizu *et al.*, 1989; Ishida-Yamamoto *et al.*, 1991; Eady & Shimizu, 1992) which we believe is satisfactory for labelling most if not all epidermal or basement-membrane antigens. It requires the use of specialized apparatus which might deter those who can achieve comparable results using a less expensive and complicated procedure.

(i)   Optional fixation in periodate–lysine–paraformaldehyde (McLean & Nakane, 1974) for 2 h at 4 °C.

(ii)   Quench free aldehyde groups with 100 mM glycine in PBS for 1 h at 4 °C. The samples can then be kept for several days, if necessary, in 0.1 M phosphate buffer at 4 °C.

(iii)   The samples can be processed into Lowicryl K4M or LR White resin using progressive lowering of temperature during dehydration and infiltration. These methods have been fully described elsewhere (Merighi, 1992; Newman & Hobot, 1987; Haftek, Chignol & Thivolet, 1989; Eady & Shimizu, 1992).

(iv)   Alternatively, the fixed samples, or others that have not been chemically fixed can be cryoprotected by soaking in 15–20% glycerol in PBS for 1 h at 4 °C.

(v)   The samples are then rapidly frozen by plunging into liquid propane at −190 °C using the Reichert–Jung KF-80 apparatus.

(vi)   The frozen skin is subjected to freeze substitution in methanol for 48 h at −80 °C using a Reichert–Jung CS Auto apparatus.

(vii) Embed the samples in Lowicryl K11M (Chemische Worke Lowi, Waldkraiburg, Germany). Polymerize the blocks under UV radiation for 48 h at 60 °C and continue for a further 48 h at room temperature.

(viii) Semithin and ultrathin sections are cut from the Lowicryl blocks and immunolabelling can be performed on the antigens exposed at the surface of the sections.

Immunolabelling for high microscopy

Different methods are available for labelling antibodies including the use of ferritin, enzymes and colloidal gold. Although the best tried methods still use peroxidase, we tend to favour indirect immunogold procedures almost exclusively. Colloidal gold labelling can be performed in a single indirect step using immunoglobulin (IgG) conjugated to gold or with protein A or protein G techniques (Horisberger & Rosset, 1977; Roth, 1984). Initially there was concern about the quality control of IgG-gold reagents (particularly relating to the variation in sizes of the gold particles) but better and more consistent reagents are now available from commercial suppliers. We use almost exclusively IgG-gold with various sizes (5–40 nm) of gold particles, and more recently 1 nm gold which is then silver-enhanced (see below).

Semithin sections (1 μm thick) are cut from Lowicryl-embedded blocks, collected on glass slides and stained with Richardson's solution (Richardson, Jarett & Finke, 1960) or toluidine blue. Further sections are treated with 1% bovine serum albumin and 5% normal goat serum (if appropriate) in PBS, pH 7.4 for 30 min to block nonspecific staining. The sections are then incubated with a primary antibody (e.g. antikeratin) at appropriate dilution in 0.5% BSA with 0.1% gelatin in PBS, washed twice in buffer, incubated with affinity-purified biotin-conjugated goat antimouse or anti-rabbit IgG (Sera-Lab; Crawley Down, Sussex, UK) for 60 min. After further washing, the sections are incubated in 1 nm colloidal gold-conjugated streptavidin (AuroProbe One Streptavidin, Amersham International, Amersham, UK) for 60 min. After further washing in buffer and distilled water, the gold particles are 'silver enhanced' to make them visible under the microscope. This is conveniently done using the IntenSE M Silver Enhancement Kit (Amersham International). The reaction time should be controlled microscopically, and generally will be 8–12 min at room temperature.

The sections are counterstained with toluidine blue or hematoxylin, dried, mounted, and observed by light microscopy.

Immunolabelling for electron microscopy

The procedure is essentially very similar to that used on semithin sections. The use of 1 nm-gold with silver enhancement has the advantage of increasing the labelling density as well as the sensitivity of the method. Alternatively labelling with IgG-gold using particles of different sizes can be used for single or double labelling experiments.

Pre-embedding immunolabelling

The traditional immunoperoxidase methods of immunolabelling are based on pre-embedding methods, which involve the incubation of the fresh or lightly fixed specimen, either as cells or more usually, slivers or sections of tissue, with the immunoreagents, before fixation and resin embedding. This approach is especially useful for localizing antigens in the extracellular matrix, including the basement membrane zone. Our approach (Shimizu *et al.*, 1990; Eady & Shimizu, 1992) has followed the method of *en bloc* immunogold labelling described for localizing type-VII collagen in anchoring fibrils and anchoring plaques by Sakai *et al.*, (1986). Localizing antigens in the lamina lucida using the pre-embedding methods is not usually possible when larger gold particles are used, probably because of the barrier effects of the lamina densa. However, the use of immunogold silver staining with 1 nm gold probes allows satisfactory labelling of a variety of basement membrane related antigens including those in the lamina lucida (McGrath *et al.*, in preparation; McGrath *et al.*, 1994).

Controls

Control incubations should be performed as with any type of immunohisto-chemical or cytochemical study. These include the substitution for the primary antibody of nonimmune serum or medium from the myeloma cell-line.

Part IV

MARKERS OF TERMINAL DIFFERENTIATION

Introduction

Two major classes of proteins that are expressed during terminal differentiation of keratinocytes are keratins and components of the cornified envelope (for reviews see Watt, 1989; Fuchs, 1990; Eckert *et al.*, 1993). This section describes protocols for identifying the keratins of hair (see Powell & Rogers, Chapter 24) and interfollicular epidermis (see Rugg, Chapter 23) and for analyzing several aspects of envelope formation. Protein precursors of the cornified envelope, including involucrin (see Sheibani & Allen-Hoffmann, Chapter 26), are synthesized prior to envelope assembly and their presence can be detected with specific antibodies (see Rice, Chapter 25). Assembly of the envelope is catalyzed by a specific calcium-dependent transglutaminase (see Rice, Chapter 25) and premature envelope assembly can be induced by treating keratinocytes with calcium ionophores (see Rice, Chapter 25).

23 Detection and characterization of keratins by immunocytochemistry and immunoblotting

E. L. RUGG

Differential keratin expression is most commonly analyzed by (i) *in situ* antibody staining of tissue sections (immunohistochemistry) or cells in culture (immunocytochemistry), or by (ii) identifying keratins separated by polyacrylamide gel electrophoresis by their characteristic charge and size migration properties or by their antibody recognition on immuno-blotting. Immunostaining *in situ* is quick and simple and provides information on the distribution of keratins within the cell, and also on changes in keratin expression during terminal differentiation. Alternatively, the cyto-skeleton can be extracted and the components resolved by PAGE followed by immunoblotting: this type of analysis provides information on the relative proportions of different keratins present, but cannot distinguish between strong expression by a subpopulation of cells and weaker expression in all the cells. The two approaches are complementary, and should both be used to study keratin expression.

Immunocytochemical staining of keratinocytes

Immunocytochemical staining depends on the reactivity of a specific antibody with a cellular target antigen, followed by visualization of the antibody binding using a suitable detection system. The cells may be fixed or unfixed and the detection system may be directly linked to the primary antibody or linked to a secondary antibody which is specific for the primary antibody.

The purpose of fixation is to preserve the cell architecture: it is important that the fixation process does not alter the morphology of the cell or the structure of the antigen so as to impair antibody/antigen recognition. There are two classes of fixatives in general use, those based on organic solvents, which remove lipids and dehydrate the cells and those which use aldehydes to cross-link proteins. Cryofixation (freezing, for example by immersion in liquid nitrogen) is frequently used for tissue sections. Most antibodies work less well on aldehyde-fixed tissue, largely due to reduced access to the antigens or changes in epitope structure, however, the disadvantage of frozen sections is the poor morphological preservation. If chemical fixation is necessary, organic solvent based fixation is recommended for staining with antikeratin antibodies.

The specificity of immunocytochemical staining is dependent on the specificity of the primary antibody. Over recent years, a large number of monoclonal antibodies have been produced which recognize specific keratins or groups of keratins; many of these are now very well characterized and an increasing number are commercially available. A list of many of these antibodies and their specificities is provided by Lane & Alexander, 1990.

Direct or indirect labelling of the primary antibody.

Direct linking of a detection system to the primary antibody gives very good resolution and low background staining. This method is particularly useful for double labelling experiments as it overcomes the requirements for secondary antibodies from different species, and for elaborate staining protocols. The major disadvantages are the reduced staining intensity compared with indirect immunofluorescence, and the time required for purification and labelling of the reagent. Commercially produced fluorescein-labelled keratin monoclonal antibodies have recently become available which may prove useful.

More commonly, the staining procedure involves the use of a labelled secondary antibody layer which is specific for the primary antibody. For example, the binding of a mouse monoclonal antibody may be detected using rabbit antiserum to mouse immunoglobulins, in which the rabbit antiserum is coupled to a detection system. This indirect method offers significant advantages over direct labelling. High-quality labelled secondary antibodies are widely available from commercial sources and the same antibody can be used to detect many primary antibodies. It also tends to produce a stronger signal due to the extra amplification level, although this may be at the cost of increased background staining.

Visualization of antibody binding

There are many systems described in the literature for the visualization of antibody binding, including the linking of antibodies to fluorochromes, enzymes, biotin, or gold particles (Polak & Van Noorden, 1987; Harlow & Lane, 1988 and references therein). These systems are all commercially available.

The fluorochromes, fluorescein, rhodamine and more recently Texas red, are commonly used in immunocytochemistry and give very good resolution. Fluorescein gives the strongest signal, but is prone to photobleaching, whereas rhodamine and Texas red are weaker but do not fade as much. The major disadvantage is that they require special optics and lighting for viewing. Fluorochromes are not generally compatible with counterstaining, as many of the commonly used stains autofluoresce. Fluorescein with either rhodamine or Texas red can be used for double labelling experiments.

Enzyme-linked antibodies are also widely used for the detection of antibody binding. Horseradish peroxidase (HRP) and alkaline phosphatase (AP) are the most commonly used enzymes, although β-galactosidase can also be used. Binding is visualized using a chromogenic substrate. Enzyme-based detection systems are extremely sensitive and do not require special optics or lighting for viewing but they do not give as good resolution as the fluorochromes. They are compatible with many commonly used counterstains.

A very sensitive method of immunostaining has been developed based on the interaction between biotin and avidin, (Hsu, Raine & Fanger, 1981). In this case, the primary antibody binds a biotinylated second antibody which in turn interacts with an avidin/biotin complex (ABC system) in which the biotin has been labelled with HRP. In this way, the primary antibody becomes labelled with several HRP molecules producing a large amplification of the original signal.

Detection systems using gold particles were originally developed for electron microscopy, and have not been widely used for light microscopy because of their lack of sensitivity compared to other methods. They do, however, give better resolution than the enzyme based methods, and sensitivity can be increased by silver enhancement.

Whichever method is used, it is important to carry out appropriate controls. For monoclonal antibodies against keratins, they would be nonkeratin monoclonal antibodies, from the same source, i.e. hybridoma supernatant, ascites, etc. preferably of the same class and subclass as the keratin antibody. Secondary antibodies should always be tested in the absence of primary antibody and when using enzyme based detection systems the cells/tissue should be checked for endogenous enzyme activity by carrying out the enzyme reaction in the absence of any antibodies.

Protocols for the immunocytochemical staining of keratinocyte cultures

Reagents

FIXATIVE: METHANOL/ACETONE (1:1)

ANTIBODY DILUTING SOLUTION: Phosphate buffered saline (PBS) containing 1% BSA or, if available, expired tissue culture medium containing 10% fetal calf serum. These solutions should be used for all antibody dilutions.

PRIMARY ANTIBODIES: The correct dilution of primary antibodies should be determined empirically. However, as a rough guide, hybridoma supernatants should be used neat, and purified monoclonal antibodies from ascitic fluid and polyclonal antibodies at concentrations of 0.1–10 μg/ml.

If the concentration of the specific antibody is unknown, test a range of dilutions from 1:10 to 1:10 000. If the antibody has been obtained from a commercial source, refer to the data sheet for guidance. Diluted antibody solutions can be stored at 4 °C, in the presence of 0.05% sodium azide.

SECONDARY ANTIBODIES: Labelled antiserum should be diluted as detailed by the supplier, and this is normally in the range of 1:10 to 1:300. Fluorochrome labelled antibodies should be stored in the dark at 4 °C to minimize photobleaching of the reagents. *A precipitate sometimes forms during storage, which may result in 'spots' of highly fluorescent material appearing on the specimen; this can be removed by filtering through a 0.22 μm filter. The filter must be prewashed with the antibody diluting solution to prevent the antibody binding to the filter. An alternative is to spin the stock solution at 10 000 rpm for 1 min. This reduces the risk of losing antibody through filter binding but it is not as effective as filtration for removing the precipitate.*

Antibodies conjugated to enzymes should be treated in a similar manner but cannot be stored. Azide inhibits peroxidase catalysed reactions and should not be included in any HRP containing solution.

MOUNTING MEDIA: There are several mounting media commercially available and it is important to establish that the one chosen is compatible with the particular method of staining employed. For example, DPX cannot be used with fluorochromes or where reaction products are alcohol soluble. The following mounting medium is suitable for most purposes including fluorochrome-stained samples (Heimer & Taylor, 1974).

- Gelvatol 20–30   2.4 g
- or Moviol 4–88
- Glycerol   6 g
- H_2O 6 ml
- 0.2 M Tris-HCl, pH 8.5   12 ml

Mix the gelvatol 20–30 (Monsanto Chemicals) or Moviol 4-88 (Hoechst) with the glycerol and add the water. Leave at room temperature for several hours. Add the Tris-HCl buffer and heat to 50 °C for 10 min with occasional mixing. When the gelvatol has dissolved, clarify by centrifuging at 5000 g for 15 min. If the mounting medium is to be used with fluorochromes, an antifade agent, such as 2.5% 1,4-diazobicyclo-[2.2.2]-octane (DABCO), should be included. Aliquot into microfuge tubes and store at −20 °C. The mounting medium is stable at room temperature for several weeks after thawing.

Fixation

The following method describes the fixation of cells grown on a 5 cm plastic tissue culture plate (Lane & Lane, 1981). This protocol can be adapted for cells grown on glass slides or coverslips. Once fixed, the cells can be stored at −70 °C for future use. The procedure should be carried out at room temperature in a fume cupboard.

(i) Decant culture medium from plate.
(ii) Flood plate with 2–3 ml of methanol/acetone and pour off.
(iii) Repeat Step (ii), but this time leave the methanol/acetone on the plate for 5 min.
(iv) Air dry in fume cupboard (approximately 30 min).

Cell staining with fluorescent antibodies

(i) Place a coverslip on the underside of a tissue culture plate and draw round it with a marker pen. Remove the coverslip and divide the area into squares (Fig. 23.1).
(ii) Very carefully load a drop (5 to 10 μl) of the primary antibody on to each square. The solution should sit as a bead on the dish (Fig. 23.1). Place the culture dish in a humid chamber (i.e. a sandwich box lined with wet filter paper) and leave at room temperature for 60 min. *If the plate is not completely dry at this stage the drops will tend to spread and mix with adjacent ones.*
(iii) Quickly pour off the primary antibody, and wash the plate under gently running tap water for 5 min. Drain well.

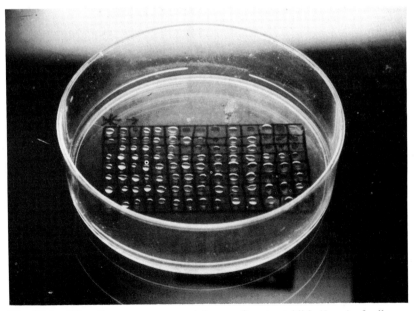

Fig. 23.1. Microfluorescence assay. A just-confluent petridish (5 cm) of cells, fixed with acetone/methanol, divided into squares and loaded with primary antibodies. Note how the antibodies sit as 'beads' and do not spread into adjacent squares.

(iv)   Cover the whole surface of the dish with secondary antibody and incubate in a humid chamber for 60 min.

(v)   Pour off the secondary antibody and rinse as described in Step (iii). Drain well.

(vi)   Add a small drop of mounting medium to the dish, then gently lower the coverslip over the cells, being careful not to trap air bubbles. Remove excess medium with a paper towel and leave to set. *Not all mounting media set completely. If the one you have used does not, seal the coverslip with clear nail varnish.*

(vii)   Cut away the sides of the petri dish with sharp scissors, and tape the bottom of the petri dish to a large glass slide.

(viii)   View sample using a fluorescent microscope equipped with the appropriate filters.

Cell staining with HRP-labelled secondary antibody

The protocol for staining cells is essentially the same as that using fluorochrome-labelled antibodies except that, with enzyme conjugated secondary antibodies, an additional reaction with a chromogenic substrate is required. The method described here uses an HRP-linked secondary antibody. This method is compatible with counterstaining with Harris's hematoxylin. For detection of alkaline phosphate- or β-galactosidase-linked antibodies, see Harlow and Lane, 1988.

Reagents

TRIS BUFFERED SALINE (TBS)

- 50 mM Tris-base   6.055 g/l
- 150 mM NaCl   8.76 g/l
- Adjust pH to 7.4 with HCl

BLOCKING SOLUTION:

- TBS   20 ml
- 10% Tween-20   0.1 ml
- BSA   0.1 g

DIAMINOBENZIDINE (DAB) SOLUTION:

- TBS   15 ml
- DAB   10 mg
- 30% H_2O_2 12 μl
- 1% $NiSO_4$ (aqueous)   150 μl
- Mix just before use

Note: DAB is a potential carcinogen and should be handled with care.

Method

(i)   Follow the protocol describing staining with fluorochrome-labelled secondary antibodies up to Step (iv).

(v)   Add 5 ml of blocking solution to the plate, and incubate at room temperature for 10 min.

(vi)   Rinse in gently running tap water for 5 min.

(vii)   Add 5 ml of DAB solution to the plate and incubate for 2–10 min. The reaction can be monitored under a microscope.

(viii)   Rinse with tap water for 5 min, followed by distilled water for 1 min.

(ix)   Counterstain with Harris's hematoxylin if required.

(x)   Apply coverslip as described earlier.

Counterstaining

This method is not suitable for use with fluorochrome-labelled specimens or where enzyme reaction products are soluble in alcohol.

(i)   Wash the specimen gently in water.

(ii)   Add a few drops of Harris's hematoxylin to the specimen and incubate for about 5 min. The time will vary depending on the intensity of staining required.

(iii)   Wash gently in water.

(iv)   Dip the specimen in 0.5% glacial acetic acid/99.5% ethanol for 10 s.

(v)   Wash gently in water.

(vi)   Dehydrate samples by passing through 75%, 95%, and absolute ethanol (3 min in each). Place for a further minute in fresh absolute ethanol and air-dry. *Alternatively, air-dry directly without passing through ethanol.*

(vii)   Apply a small drop of DPX to the specimen, and carefully lower coverslip avoiding trapping air bubbles. Leave to set.

Troubleshooting

The most commonly encountered problem is high background staining. This can arise from two sources, cross-reactivity with antigens in the specimen or general nonspecific binding not involving the antigen binding site. Specific cross reactivity, especially when using monoclonal antibodies, may be due to sequence homology between two antigens and may only be resolved by using another antibody. Specific cross reactivity resulting from secondary antibody interactions may be resolved by incubating the specimen with a solution containing 1% serum from the same species as the second antibody.

Polyacrylamide gel electrophoresis and immunoblotting

Proteins can be separated according to their size by SDS–polyacrylamide gel electrophoresis (SDS–PAGE), transferred from the gel to a membrane and then probed with antibodies to specific proteins (immunoblotting). Human keratins range in size between 40 and 68 kDa and, since several migrate at similar rates on SDS–PAGE gel, it is difficult to resolve individual proteins. To overcome this difficulty, cytoskeletal extracts are often analyzed by two-dimensional electrophoresis, a method which takes advantage of the large charge differences between type I (basic) and type II (acidic) keratins. First the keratins are separated according to their charge, then according to size and, finally, they are analyzed by immunoblotting. Two-dimensional electrophoresis has, to some extent, been superseded by the wide availability of monoclonal antibodies which are specific for individual keratins and allow a complex mixture of keratins to by analysed by SDS–PAGE and immunoblotting alone.

Choice of gel concentration and degree of cross-linking

Polyacrylamide gels are formed by the cross-linking of acrylamide monomers with N' N' methylene bisacrylamide, to form a matrix of approximately uniform pore size. Polymerization is initiated by free radicals which are generated by ammonium persulphate, the formation of free radicals being catalyzed by the inclusion of N,N,N',N'-tetramethylethylenediamine (TEMED) in the reaction. Within certain limits the rate at which a given protein will migrate through an acrylamide gel is proportional to the size of the pores, i.e. a combination of the concentration of acrylamide and the degree of cross-linking. The total gel concentration (T) and amount of crosslinking (C) can be determined from the following formula:

$$\% T = \frac{\text{Acrylamide (g)} + \text{Bisacrylamide (g)}}{\text{Volume}} \times 100$$

$$\% C = \frac{\text{Bisacrylamide (g)}}{\text{Acrylamide (g)} + \text{Bisacrylamide (g)}} \times 100$$

For the separation of keratins extracted from keratinocyte cultures, total acrylamide concentrations of between 10% and 12.5%, with 2% to 2.8% cross-linking are recommended.

Molecular standards

A set of standard proteins for the determination of molecular weight should be included in all electrophoretic analyses. Prestained molecular standards are commercially available and have the advantage that they can be used as a measure of protein transfer during the blotting procedure. However, the migration rate of these prestained proteins may be different to their unstained counterparts, and care is needed when interpreting results. A more appropriate set of molecular weight markers for keratins are cytoskeletal extracts from well-characterized cell lines which express keratins.

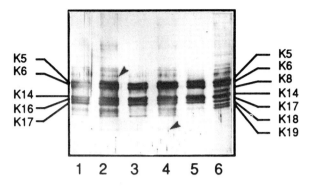

Fig. 23.2. A blot of a SDS–PAGE gel, stained with India Ink. lanes 1 to 5 show the cytoskeletal extract obtained from five different primary human keratinocyte cultures; lane 6 is a similar extract obtained from TR146 cells. The 'fuzzy' band (top arrow) indicates contamination of the sample with keratin 1 derived fragments which make up a large proportion of laboratory dust. The 'streaks' (lower arrow) are due to dust particles on the glass plates.

For example, the cytoskeletal extract from TR146 cells contains large amounts of K5, K6, K14, K17, and smaller but detectable quantities of, K8, K18 and K19 (Rupniak *et al.*, 1985; Fig. 23.2). The extract from these cells provides an excellent set of molecular weight markers which also serve as controls for the subsequent immunostaining.

Electrophoresis Apparatus

The particular electrophoresis system used depends on the needs of the experimenter. For most applications, a slab gel apparatus which gives minigels of about 8 × 8 cm is sufficient; however, if better resolution is required longer gels should be run. When running longer gels, heating may become a problem, negating the extra resolution from the increased length. Some apparatuses have integral cooling systems, others may need to be run in a coldroom. Many of the slab gel apparatuses can be adapted to run two-dimensional gels.

Membranes for blotting

Polyacrylamide gels cannot be directly probed with antibodies and it is therefore necessary to transfer the proteins from the gel to a membrane compatible with immunostaining. Nitrocellulose (0.22 µm) is the most commonly used membrane; however, activated nylon membranes can also be used. Nylon membranes have the advantage of being less brittle than nitrocellulose, although they do tend to give higher backgrounds.

Monitoring transfer

The degree of transfer from gel to membrane can be monitored by the use of prestained molecular weight markers, or the addition of a suitable dye to the sample. One such dye is pyronin Y which migrates just in front of bromophenol blue on the gel. Alternatively, after transfer, the blot can be stained for protein with Ponceau S. This is a reversible protein dye which can be washed from the blot and does not interfere with the subsequent reactions.

Staining the immunoblot

Once the proteins have been transferred, the membrane is blocked to reduce the nonspecific binding and then probed with antibodies specific for the protein of interest. The detection of antibody binding is achieved using an enzyme-linked secondary antibody in a manner similar to that for immunostaining of cells; horseradish peroxidase- and alkaline phosphatase-linked secondary antibodies are commonly used and their substrates may be chromogenic or light emitting.

Extraction of keratins

Keratins are detergent-insoluble proteins, and this property is used to produce cytoskeletal extracts from cells which are enriched with keratins. The following protocol describes the extraction of keratins from keratinocyte cultures (Stasiak *et al.*, 1989), although the method can be adapted to extract keratins from tissue. In this case, the tissue should be frozen in liquid nitrogen and ground to a fine powder with a mortar and pestle before processing.

Reagents

0.01 M PHOSPHATE BUFFERED SALINE (PBS):

- Na_2HPO_4 2.047 g/l
- $NaH_2PO_4.2H_2O$ 0.257 g/l
- NaCl   8.77 g/l

LOW SALT EXTRACTION BUFFER:

- 10 mM Tris-base   1.21 g/l
- 150 mM NaCl   8.77 g/l
- 3 mM EDTA   0.12 g/l
- 0.1% N-P40   1.00 g/l

Dissolve salts in distilled water and adjust pH to 7.4 with HCl.

HIGH SALT EXTRACTION BUFFER:

- 10 mM Tris-base   1.21 g/l
- 150 mM NaCl   8.77 g/l
- 1.5 M KCl   111.8 g/l
- 3 mM EDTA   0.12 g/l
- 0.1% N-P40   1.00 g/l

Dissolve salts in distilled water and adjust pH to 7.4 with HCl.

WASH BUFFER:

- 10 mM Tris-base   1.21 g/l
- 150 mM NaCl   8.77 g/l
- 3 mM EDTA   0.12 g/l

Dissolve salts in distilled water and adjust pH to 7.4 with HCl.

Keratins are relatively resistant to protease attack; however, as a precaution against degradation it is advisable to add protease inhibitors to the extraction and wash buffers. The inclusion of 0.2 mM PMSF, 0.5 μg/ml leupeptin and 0.5 μg/ml pepstatin will inhibit most proteases.

100 mM PMSF: 17.4 mg/ml dissolved in isopropanol. This solution is stable at room temperature for several months; however, once diluted in aqueous solution, it has an half-life of about 30 min. It therefore should be added to solutions just before they are required.

LEUPEPTIN: 1 mg/ml dissolved in distilled water. This solution can be kept frozen in small aliquots. Once thawed, it can be stored at 4 °C for up to 1 month.

PEPSTATIN: 1 mg/ml dissolved in distilled water. Pepstatin is stable at room temperature for several months.

Method

The extraction of keratins from keratinocyte cultures grown on a 5 cm tissue culture dish. After the initial washes the procedure should be carried out at 4 °C to minimize the risk of protein degradation.

(i)   Gently pour off culture medium.
(ii)   Rinse cells with 3 × 5 ml of warm PBS or serum free tissue culture medium (37 °C).
(iii) Add 5 ml of low salt extraction buffer and leave 10 min on ice.
(iv)   Scrape cells and extraction buffer into a 15 ml plastic centrifuge tube. Spin for 10 min at 2000 rpm. Discard the supernatant.
(v)   Add 5 ml of high salt extraction buffer to the pellet, and leave 10 min on ice.

(vi)   Spin for 10 min at 2000 rpm. Discard the supernatant.

(vii)   Wash pellet with 15 ml of wash buffer. *This is best achieved by adding approximately 3 ml of wash buffer, vortexing vigorously until the pellet is dispersed and then making up to volume.*

(viii)   Spin 15 min at 2000 rpm. Discard supernatant.

(ix)   Repeat Steps (vii) and (viii) once. *The pellet can be extremely 'sticky' and difficult to disperse at this stage. Extensive vortexing may be required. Sonication can help.*

(x)   Resuspend the final pellet, which is predominantly keratins, in 0.5–1.0 ml of either SDS–PAGE or NEPHGE sample buffer. *A significant proportion of the cytoskeletal extract will be deposited over the inside surface of the tube. To maximize the recovery of sample gently roll the sample buffer around the sides of the tube. A very viscous solution at this stage is due to the presence of contaminating DNA which can lead to difficulty when loading samples onto gels. The DNA can be sheared by passage through a 19 ga syringe needle 4–6 times, or by sonication. Remember SDS–PAGE sample buffer contains detergent and so try to avoid excessive frothing. The addition of DNase (25 μg/ml) to the sample will also help.*

(xi)   Transfer to 1.5 ml Eppendorf tubes and then (SDS–PAGE samples) boil for 5 min in a boiling water bath (a pinhole in the lid will prevent it popping off with possible loss of sample) or (NEPHGE samples) heat to 50 °C for 5 min.

(xii)   Store the samples in 50 or 100 μl aliquot at −70 °C. They are stable for at least 1 year.

SDS–Polyacrylamide gel electrophoresis (SDS–PAGE)

The method described here is for a discontinuous slab gel as originally reported by Laemmli, 1970. The proteins in the sample are first concentrated into zones in the stacking gel, and then separated according to size in the resolving gel. This system has the advantage that relatively large sample volumes can be loaded without significant loss of resolution.

Preparation and assembly of plates

The assembly of plates to form a cassette depends on the particular apparatus. Methods vary from the use of sticky tape to sophisticated clamping systems. The object of all these methods is to form a sealed system in which the gel can polymerize, and to hold the gel during electrophoresis. It is imperative that all the components of the system are scrupulously clean. This is particularly important when working with keratins, since a large amount of the dust in the air is derived from skin. Contamination with dust severely affects the quality of the final gel (see Fig. 23.2); gloves should be worn throughout to avoid contamination with hand proteins.

Reagents

STOCKS:

- 30% w/v acrylamide
- 2% w/v bisacrylamide
- 10% w/v SDS
- TEMED
- 10% w/v ammonium persulphate (APS)
- × 4 resolving gel buffer   1.5 M Tris-HCl, pH 8.8

- × 4 stacking buffer   0.5 M Tris-HCl, pH 6.8   1

- × 10 running buffer   0.25 M Tris-base   30.3 g/l
-         1.92 M glycine   144 g/l
-        1% SDS   10 g/l

SDS SAMPLE BUFFER:

- 2% w/v SDS   0.2 g
- 10% w/v glycerol   1 g
- 100 mM DTT   0.15 g
- × 4 stacking buffer   1.25 ml
- 0.02% w/v bromophenol blue   2.0 mg

Make up 10 ml with distilled water. Store at $-20°$ in small aliquots

The quality and resolution of gels are affected by the purity of the reagents, and where possible electrophoresis grade reagents should be used. Solutions of acrylamide and bisacrylamide are available from commercial sources and minimize the risk of exposure to acrylamide which is a neurotoxin. Batch variability can be a problem with commercially obtained stocks and may affect resolution.

Preparation of gel cassette

Wash the glass plates in detergent and rinse thoroughly with running tap water followed by distilled water, then methanol. Air dry. Lay the plates on a clean surface and give the upper surface a final clean with methanol. Assemble the cassette according to the manufacturers' instructions.

Preparation of gels

RESOLVING GEL: Makes five to eight 8 × 8 cm minigels.
$(T = 10\%, C = 2\%)$:

- 30% w/v acrylamide   16.35 ml
- 2% w/v bisacrylamide   5.00 ml

- × 4 resolving gel buffer   12.50 ml
- 10% SDS   0.50 ml
- Distilled H_2O 15.25 ml
- TEMED   0.025 ml
- 10% APS   0.375 ml

Thoroughly mix the first five solutions, then add the TEMED and freshly prepared APS. Pipette the solution into prepared cassettes. *Oxygen inhibits polymerisation so try to avoid airation of the gel solution during the preparation.* Carefully overlay the upper surface of the gel with 80% isopropanol or diluted resolving gel buffer. This excludes oxygen and produces a level gel surface. Allow gel to polymerize (approximately 1 h, depending on the ambient temperature). Several gels may be prepared at the same time and stored for up to 1 week at 4 °C, wrapped in wet paper towels and saran wrap. Do not add the stacking gel until just before use.

STACKING GEL: The stacking gel should have a relatively large pore size, and it should be at least twice sample volume.

$(T = 5\%, C = 2\%)$
- 30% acrylamide   0.82 ml
- 2% bisacrylamide   0.25 ml
- × 4 stacking gel buffer   1.25 ml
- 10% SDS   0.05 ml
- Distilled H_2O 2.58 ml
- TEMED   0.005 ml
- 10% APS   0.0375 ml

Remove the overlay and any unpolymerized acrylamide from the top of the resolving gel by washing with distilled water. Drain well and remove any excess water with blotting paper, being careful not to touch the gel surface.

Make up the stacking gel by mixing the first 5 solutions. Add the TEMED and freshly prepared APS just before use. Pipette stacking gel on to the top of the resolving gel and insert comb avoiding trapping any air bubbles. Allow to polymerize. This will take approximately 10–15 min depending on the ambient temperature.

Assemble cassette containing polymerized gel into the electrophoresis apparatus and fill the upper and lower reservoirs with running buffer (× 1), making sure that there are no air bubbles trapped at the top or bottom of the gel. Wash the sample wells with running buffer to remove any unpolymerized acrylamide.

Sample loading and running gel

Spin samples in a microcentrifuge for 3 min at 12 000 rpm, to remove any undissolved material. Very carefully, layer the supernatant into the sample wells using a micropipette fitted with a tip which will reach the bottom of the well. The exact electrophoretic conditions will depend on the composition and dimensions of the gel and should be determined empirically. As a rough guide, an 8 × 8 cm gel should be run at 25 to 30 mA until the bromophenol blue dye front reaches the bottom of the gel. This will take approximately 60 min.

Two-dimensional gel electrophoresis

The method described here is based on that reported by O'Farrell, Goodman & O'Farrell, 1977, which describes the separation of basic and acid proteins using nonequilibrium pH gradient gel electrophoresis (NEPHGE). The proteins are initially separated on a rod gel according to their charge, followed by separation according to size using SDS–PAGE and finally they are analyzed by immunoblotting (Figs. 23.3 and 23.4). The pH gradient in the first dimension is generated by the inclusion of ampholines in the gel. Ampholines vary greatly between batches, and this

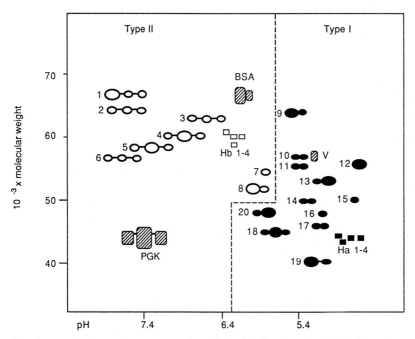

Fig. 23.3. A schematic representation of the distribution of epithelial keratins (1-20) and trichocytic (Ha 1-4 and Hb 1-4) following analysis by two-dimensional electrophoresis (data compiled from Moll *et al.*, 1990, O'Guin *et al.*, 1990, and personal observations). The positions of commonly used markers, bovine serum albumin (BSA), vimentin (V), and 3-phosphoglycerate kinase (PGK) are also shown.

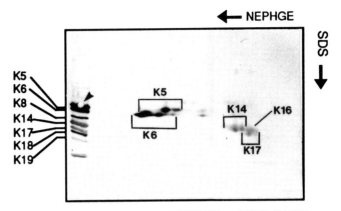

Fig. 23.4. The blot of a two-dimensional (NEPHGE and SDS–PAGE) analysis of a cytoskeletal extract obtained from a primary culture of human keratinocytes. The blot was first stained with αIFA antibody which recognizes all intermediate filaments (Pruss *et al.*, 1981), although with varying affinities, and then stained for total protein with India ink. The more heavily stained bands are the result of stronger immunostaining. A TR146 cytoskeletal extract was included as a molecular weight marker for the SDS–PAGE analysis (left-hand lane). The arrow indicates the effect of a bubble between the gel and the membrane during electrophoretic transfer.

affects both reproducibility and resolution. To make comparisons between gels, the same batch should always be used.

Preparation of tubes for rod gel
Clean with detergent and rinse thoroughly with tap water, followed by 5 M HCl. Finally rinse with 0.1 M KOH in 50% ethanol and dry. Seal one end of each tube with parafilm.

Reagents

STOCK SOLUTIONS:

- 30% w/v acrylamide
- 2% w/v bisacrylamide
- 10% w/v N-P40

AMPHOLINES:

- pH 3.5–10
- pH 5–8

UPPER ELECTRODE BUFFER:

- 0.085% H_2PO_4

LOWER ELECTRODE BUFFER:

- 0.02 M NaOH

NEPHGE SAMPLE BUFFER:

Urea	5.83 g
DTT	0.15 g
Ampholines (2%)	
pH 5–8	0.16 ml
pH 3.5–10	0.04 ml
10% w/v N-P40	2.00 ml
Lcupeptin	0.20 mg

Make up to 10 ml with distilled water.

SDS EQUILIBRATION BUFFER:

• 0.5 M Tris-HCl, pH 6.8	12.5 ml
• SDS	2.3 g
• Glycerol	10 ml
• Bromophenol blue	0.02 g

make up to 100 ml with distilled water.

Preparing gels

Urea	1.1 g
30% w/v acrylamide	0.25 ml
2% w/v bisacrylamide	0.21 ml
10% w/v N-P40	0.4 ml
Ampholines (2%)	
pH 3.5–10	0.15 ml
pH 5–8	0.05 ml
Distilled H_2O	0.2 ml

Combine the above reagents and heat to 37 °C, swirling gently to mix. Filter through a 0.22 μm filter. Add 2 μl TEMED and 2 μl 10% APS.

Fill the tubes, to the top, using a 1 ml syringe fitted with plastic tubing which reaches the bottom of the tube. *Be careful to avoid introducing an air bubble at the bottom of the gel. If this occurs, it can be removed by gently bouncing the tubing on the parafilm.*

Remove 20 μl from the top of each tube and overlay with water. Cover to prevent evaporation. Leave at least 2 h (or overnight) to polymerize. *If the urea crystallizes out, it can be redissolved by warming with a lamp or over a water bath.*

Electrophoresis

Flick the water from the top of each tube and place in the electrophoresis apparatus. Thaw out the sample at 37 °C to dissolve the urea, and then microfuge for 1 min at 14 000 rpm. Load the sample immediately. Place the appropriate electrode buffer in the upper and lower chambers of the electrophoresis apparatus, and connect the electrodes in REVERSE POLARITY. Run for 3 min at 200 V, 3 min at 300 V, followed by 2.5 h at 400 V (1 kV h).

Removal and equilibration of gels

Attach a 20 ml syringe filled with water to the top of the tube and extrude the gel into a dish. Note the orientation of the gel. Equilibrate for 5 min in SDS equilibration buffer. The NEPHGE gel can be stored at −20 °C for at least 1 year.

Second dimension (SDS–PAGE)

Electrophoresis in the second dimension is essentially as described earlier. The tube gel is placed directly on the top of a suitable SDS slab gel (e.g. 10% *T*, 2% *C*) and run as before. If necessary the rod gel may be held in place with 1% agarose. *Care must be taken at this stage not to stretch the gel when placing it on the top of the slab gel. Remember to note the orientation of the rod gel.*

Immunoblotting

Reagents

BLOCKING SOLUTION:

- PBS, 0.05% Tween 20
  or
- PBS, 5% dried milk protein (e.g. marvel)

BLOTTING BUFFER:

- Tris-base   30.3 g
- Glycine   144 g
- Methanol   1.25 l

Dissolve the tris-base and glycine in distilled water and make up to 3.75 l. Add methanol.

Method

All procedures should be carried out wearing gloves.

(i)   Cut a piece of membrane and 4 pieces of Whatman 3MM paper to the size of the gel and place in blotting buffer. *Some membranes, require prewetting in methanol before use. Check manufacturers' instructions.*

(ii)   Carefully disassemble the gel cassette, leaving the gel attached to one of the glass plates. *Make sure you know the orientation of the gel. Work quickly from now on to avoid the gel drying out.*

(iii) Assemble the blotting cassette as follows, in a tray containing blotting buffer.

- Bottom of cassette
- 1 porous support pad
- 2 pieces of Whatman 3MM paper
- the polyacrylamide gel
- 1 piece of membrane (mark a corner to indicate the orientation of the gel)
- 2 pieces of Whatman 3MM paper
- 1 porous support pad
- Top of cassette

When assembling the cassette, it is important that no air bubbles are trapped as these will interfere with the flow of current and prevent the transfer of protein (Fig. 23.4). Rolling a pipette firmly over the top of the sandwich before the cassette is closed can help to dislodge any bubbles which may have become trapped.

(iv) Place the assembled cassette in a blotting tank filled with blotting buffer, with the membrane towards the positive electrode. The proteins in the gel carry a net negative charge and migrate toward the positive electrode.

(v) Run at 200 mA for 1 h.

(vi) Disassemble the sandwich and retrieve the membrane sheet. Successful transfer can be seen by the presence of the prestained standards or pyronin Y on the membrane. Not all the protein will have been transferred in this time, but sufficient will have been for immunostaining. The gel can be stained for remaining protein with Coomassie blue.

(vii) The blot may be stained at this stage with Ponceau S.

(viii) Place in 10 to 15 ml of PBS/Tween 20 and leave to shake gently overnight or in PBS/dried milk for 1–2 h. *The choice of blocking reagent depends on how the blot is to be further processed. Blocking with dried milk protein is not compatible with staining for total protein at a later stage.*

Processing the blots to visualize the keratin

The blocked membrane is incubated with the primary antibody, washed, and then incubated with a secondary antibody conjugated to a suitable detection system, in this case alkaline phosphatase. In this reaction, the substrate 5-bromo-4-chloro-3-indoyl phosphate (BCIP) is converted into a compound which reacts with nitro blue tetrazolium (NBT) to generate an insoluble dark blue dye deposit. The dilution of the both the primary and secondary antibody should be determined empirically. However a good starting point for hybridomas supernatants is 1:10, and for enzyme linked anti-immunoglobulins 1:500 to 1:1000. The specificity of the reaction should be determined by incubation of a blot with the secondary antibody alone.

Reagents

ANTIBODIES: These should be diluted and stored as described for immuno-cytochemical cell staining.

TBST BUFFER:

- 10 mM Tris-HCl (pH 8.0)
- 0.15 M NaCl
- 0.5% Tween 20

AP BUFFER:

- 0.1 M Tris-HCl (pH 9.5)
- 0.1 M NaCl
- 5 mM MgCl$_2$

NBT: 10 mg/ml nitro blue tetrazolium in water.

BCIP: 50 mg/ml BCIP in 100% dimethyl formamide.

Method

(i)   Rinse blot with tap water and incubate with primary antibody for 1 h at room temperature on a rotary shaker. The amount of antibody required is dependent on the size of the blot. 5 ml is sufficient for a 8 × 8 cm blot in a 10 × 10 cm petri dish.

(ii)   Decant the primary antibody solution into a storage vessel. *The primary antibody can be used several times, providing it is not contaminated.* Place the petri dish containing the blot under a gentle stream of running cold tap water and wash for 5 min. Drain the petri-dish and blot.

(iii) Immerse the blot in 5 ml secondary antibody solution and shake gently for 1 h at room temperature.

(iv)   Discard the secondary antibody and wash with tap water as before.

(v)   Incubate blot in TBST for 15 min.

(vi)   Make up substrate by adding 330 μl NBT and 33 μl BCIP to 10 ml AP buffer.

(vii) Wash the blot briefly in tap water.

(viii) Place the blot, protein side up, in the substrate solution and shake gently by hand until the color develops (2–10 min).

(ix)   Wash briefly in tap water to stop the reaction and air dry on blotting paper.

Troubleshooting

Most problems encountered in immunoblotting arise from either the cross-reactivity of the antibody with other antigens (specific background staining) or diffuse staining of the whole blot (nonspecific background staining). Whether the background staining is a result of the primary or second antibody can be determined from incubation of the blot with the second antibody alone. If the background staining results from the second antibody, try using an alternative second antibody. If the staining persists, try adsorbing the secondary reagent against a preparation of the antigen. Nonspecific background staining is often a result of insufficient blocking of the membrane so try a different blocking buffer and/or adding detergent to the antibody and washing solutions. Reducing the length of incubation with the substrate or using less sensitive substrates may help. Specific background staining resulting from cross-reaction of the primary antibody may be due to a common epitope. If this is the case, it can only be resolved by the use of another primary antibody.

Protein staining

India ink

The blot can be stained for total protein with India ink solution. Staining can be carried out on probed or unprobed blots.

Reagents

INDIA INK SOLUTION: 100 μl India ink in 100 ml of 0.3% Tween-20/PBS. *Note*: The type of India ink is important. Pelikan Fount India drawing ink (black) or its equivalent is suitable.

Method
(i)   Incubate blot in blocking solution for at least 15 min.
(ii) Wash briefly with tap water.
(iii) Add 5 ml India ink solution and incubate for 0.5 to 18 h.
(iv) Stop incubation when suitably stained. Wash with tap water and dry carefully on blotting paper.

Ponceau S Staining

Reagents
2% Ponceau S in 30% w/v trichloroacetic acid.
Dilute 1:10 in distilled water for use.
The diluted solution can be reused several times.

Method
(i)   Rinse the blot with distilled water followed by a small volume of diluted Ponceau S.

(ii) Cover the blot with Ponceau S and incubate gently shaking for 5 min.

(iii) Pour off the dye solution and rinse with distilled water. The Ponceau S is removed preferentially from the membrane, leaving the protein stained red. *This step needs to be carefully monitored, as too much washing will result in total destaining of the blot.*

Coomassie blue staining

Reagents

STAIN: 0.25% Coomassie brilliant blue (R-250)
       Dissolved in 50% methanol, 10% acetic acid.

DESTAIN: 5% methanol, 7.5% acetic acid.

Method

(i) Place gel in a clean container, and add approximately 5 volumes of staining solution. Incubate 2 to 12 h, shaking gently.

(ii) Decant the staining solution into a container, and replace with destain. *The stain can be reused several times.* Incubate at room temperature, shaking gently, replacing the destain as required, until the gel is suitably destained.

Staining and destaining times can be reduced by heating the solutions to 50 °C. The inclusion of some dye absorbing material, such as a paper tissue, or a piece of foam, also reduces the time for destaining.

24   Analysis of hair follicle proteins

B. C. POWELL and G. E. ROGERS

Two-dimensional polyacrylamide gel analysis of hair keratin

The procedure outlined enables hair proteins to be displayed in two dimensions for the recognition of the different families of proteins and some individual members of the families. The technique is applicable to mature hairs and also to follicle proteins. The procedure is based on the methods described by Marshall and Gillespie (1977, 1982) and has given reproducible results in our hands in determining for example, changes in the abundance of proteins in hairs of transgenic mice and sheep compared with nontransgenic controls. However, it has to be borne in mind that the extractability of the hair proteins, even with the vigorous conditions described, varies between species. The reason for this is not known (Gillespie, 1991). The method described is to be distinguished from the extraction of hair follicle and hair keratin intermediate filament (IF) proteins by the standard IF cytoskeletal procedure and their analysis by two-dimensional electrophoresis using iso-electric focusing in one direction and SDS (sodium dodecylsulphate) in the second direction (Heid, Werner & Franke, 1986). In that procedure, only the keratin IF protein can be identified (in the sulfhydryl form) whereas the associated proteins of the keratin complex, of which there are many, such as the cysteine-rich and glycine/tyrosine-rich protein components, have not been detected.

Extraction of the protein

Washing of wool and hair
Fibers are washed with petroleum spirit $(2\times)$, ethanol $(2\times)$, water and finally ethanol again, then air dried at $20\,°C$.

Protein extraction

STOCK EXTRACTION SOLUTION:

- 8 M urea, 1 mM EDTA,
- 10 mM Tris pH 7.4.

(i)   Take 10 ml of stock extraction solution and alter pH to between 9.5 and 10.0 with 5 M KOH (requires approx. 25 μl).

Keratinocyte methods by Irene Leigh and Fiona Watt
© Cambridge University Press, 1994, pp. 149–155

(ii) To 10 mg of washed wool or hair add 1.0 ml pH 10.0 extraction solution and 30 µl 1 M dithiothreitol (DTT). Incubate at 37 °C for 16 h.

(iii) Sonicate 30 s (Branson Sonifier Cell Disruptor, B-30, set at 6).

(iv) Neutralize extract by adding 50 µl 1 M Tris pH 7.4.

S-carboxymethylation with iodo[2-^{14}C] acetic acid

Iodo [2-^{14}C] acetic acid (^{14}C-IAA) 250 µCi (9.25 MBq, 2.04 GBq/mmol) comes as a freeze dried residue in a glass ampoule so dissolve in 250 µl of water to give a 1 µCi (3.75×10^4 Bq)/µl working solution. Store at -15 °C in 50 µl aliquots.

(i) To 100 µl of neutralized extract in an Eppendorf tube, add 10 µl of 1 µCi/µl ^{14}C-IAA and leave at 20 °C for 20 min.

(ii) Then add 10 µl 1M DTT, mix and leave for 1–2 min before adding 50 µl of freshly prepared 2.3 M Tris, 1 M 'cold' (IAA) (278 mg Tris, 186 mg iodoacetic acid, water to 1.0 ml, vortex vigorously until dissolved) and leave at 20 °C for 20 min.

(iii) Finally, add 10 µl β-mercaptoethanol.

(iv) Centrifuge 2 min in Eppendorf centrifuge to pellet undissolved cellular debris.

(v) Remove supernatant and store at -15 °C.

First-dimension separation is at pH 8.9

Solutions

- A   36.6 g Tris/100 ml, adjust pH to 8.9 with concentrated (10 M) HCl.

- B   8.55 g Tris/150 ml, adjust pH to 7.0 with HCl.

- N   175.2 g acrylamide, 4.8 g N,N′-methyl-ene-bis-acrylamide (bis)/600 ml. Store at 4 °C, covered from light.

- 10 × stock electrophoresis reservoir buffer   30 g Tris, 144 g glycine/litre. (NB dilute 10 × prior to use).

- 2 × sample buffer   6 g urea, 2 ml solution B, 5 ml water, bromophenol blue to colour ($\sim$1 mg). (NB: dilute with equal vol of sample prior to loading.)

Preparation of gel:

Set up gel apparatus (Hoefer Vertical Slab Gel Unit, SE600) (glass plates, $\sim$14 cm × $\sim$14 cm × 1 mm).

SEPARATING GEL: 9.6 g urea, 2.5 ml solution A, 5 ml solution N, 5.3 ml water, warm to keep solution at 20 °C while urea dissolves. Add 100 μl 10% ammonium persulphate (APS), and 10 μl N,N,N'-tetramethylethylene diamine (TEMED) and immediately pour separating gel to a height of 12–13 cm. Overlay with water-saturated n-butanol and allow to set for 20 min. Pour off the butanol layer and proceed with the stacking gel step.

STACKING GEL: 7.7 g urea, 2 ml solution B, 2.15 ml solution N, 6.1 ml water, warm to keep solution at 20 °C as urea dissolves. Add 100 μl of 10% APS and 10 μl TEMED and immediately pour stacking gel to a height of 2.5 cm above separating gel. Allow to set. Use 1 mm 'comb' to form 0.7 cm wide wells, 1 cm into stacking gel.

SAMPLE: up to 50 μl total vol can be loaded on each well. This consists of equal volumes of 2 × sample buffer and extract.

Set up complete gel apparatus for the first dimension run in pH 8.9 gel

Remove comb from gel, fill upper (negative) and lower (positive) reservoirs with 1 × electrophoresis buffer. Load sample(s) and run at a constant current of 20 mA (set voltage limit to 400 V) until the bromophenol blue tracker dye has migrated 9 cm into the separating gel. This takes approx 3 h.

Second-dimension separation in SDS (sodium dodecylsulphate) gel

(This should follow as soon as possible after completion of the pH 8.9 slab gel run.)

Solution

● N	As for first dimension 8.9 gel.
● L	27.2 g Tris and 0.6 g SDS/150 ml, adjust to pH 8.8 with HCl.
● M	9.08 g Tris and 0.6 g SDS/150 ml, adjust to pH 6.8–7.0 with HCl.
● 10 × stock electrophoresis (reservoir) buffer	75.5 g Tris, 360 g glycine, 25 g SDS/2.5 litres water. (NB Dilute 10 × prior to use.)

Set up gel plates with 1.5 mm separation

SEPARATING GEL: 20 ml solution N, 15 ml solution L, 24.4 ml water, 400 μl 10% APS, 40 μl TEMED, pour to a height of 12.5 cm, overlay with water-saturated n-butanol as previously and allow to set.

STACKING GEL: 2.7 ml solution N, 2.0 ml solution M, 15.2 ml water, 180 μl 10% APS, 24 μl TEMED. Remove excluded liquid at top of separating gel and then pour in 1.5 cm of stacking gel solution. Allow to set.

Transfer of first-dimension to second-dimension gel

(i)   Cut sample track from pH 8.9 slab gel and equilibrate in approx. 50 ml of 1 × SDS stock electrophoresis reservoir buffer (see 'solutions' above) for 15 min immediately prior to beginning SDS run.

(ii)   After equilibration, position the pH 8.9 track on top of the SDS stacking gel.

(iii)   Run at 14 mA overnight (2 SDS gels can be run simultaneously in the Hoefer unit) and continue next morning (increasing current to 80 mA maximum to save time) until bromophenol blue marker dye has migrated 10 cm into separating gel. Stop run.

(iv)   Dismantle gel apparatus and gel plates. Mark gels for identification and orientation.

(v)   Fix gels in ~1 litre 10% ethanol, 7% acetic acid for 40 min (all equilibration, fixation or washing steps are done with gentle agitation).

(vi)   Wash gels for 15 min with 10% ethanol.

(vii)   Agitate in 1 M sodium salicylate for 30 min.

(viii) Place gels on glass sheet and overlay them with Whatman 3 MM support. Lift paper with gels adhering.

(ix)   Dry on a vacuum gel dryer.

(x)   Expose dried-down gels to X-ray film in an autoradiograph cassette with intensifying screen.

(xi)   Length of exposure depends on ^{14}C-radioactivity incorporated into protein and amount of protein loaded onto the first-dimension pH 8.9 gel.

(xii)   Develop X-ray film. If 3×10^5 cpm were loaded on the first dimension run then a satisfactory autoradiograph is obtained on an overnight exposure.

Preparation of hair fibers for transmission electron microscopy

The fine structure of mature hairs is not clearly visible without several steps of pretreatment to enable the penetration of heavy metals into the hardened structures. The most successful preparative procedure devised by Rogers (1959b) involved partial reduction of the disulfide bonds of the keratins by soaking in mercaptans (e.g. β-mercapto-ethanol) before reacting with standard osmium tetroxide fixatives. The subsequent embedding in epoxyresins and sectioning is followed by section-staining with standard uranyl acetate and lead hydroxide solutions to intensify the contrast.

Filament–matrix structures are revealed by this procedure as are inter-cellular layers and the sub-layers of the scale cells of the hair cuticle.

Wash hair fibers in petroleum ether (2 ×), ethanol (2 ×), water, ethanol and dry.

Prepare the reducing solution: 0.4 M thioglycolic acid in 0.2 M sodium acetate pH 5.5 (final).

Tie fibers into bundles with cotton thread and trim these bundles to ∼4 mm length.

Reduction
Agitate bundles of fibers in reducing solution for 24 h at 20 °C (make sure liquid is completely wetting the fibers – air bubbles trapped in bundles can be removed by applying a negative pressure on a water pump).

Washing
Remove reducing solution and wash fibers for 30 min in 0.1 M sodium phosphate buffer, pH 7.3.

Fixation
Agitate fibers in 1% OsO_4 in 0.1 M sodium phosphate buffer, pH 7.3 for 24 h.

Dehydration
Remove OsO_4 and dehydrate fibers in 50% ethanol 1 h, 70% ethanol 1 h, redistilled ethanol 2 × 1 h.

Resin infiltration
50% ethanol/50% Spurrs resin 1 h, 100% resin 2 × 1 h, 100% resin overnight, using rotational agitation to mix.

Embedding
Pick up bundles of fibers with forceps and align in embedding mold for cross-sections or longitudinal sections. Cure resin at 65 °C for 24–48 h.

Sectioning
Using a diamond knife cut sections silver to gold in color (60–100 nm) and pick up on carbon-coated parlodion 300 mesh Cu/Pd grids.

Staining
Uranyl acetate ($UrAc_2$) followed by lead citrate.

Float grids section-side down (covered from light) on 1% aqueous $UrAc_2$ containing 0.1% glacial acetic acid for 20–30 min 20 °C then wash 2 times with water.

Repeat for 4 min on Reynolds lead citrate* diluted 5 × with 0.02 M NaOH just prior to use (in CO_2-free atmosphere – NaOH pellets in a covered petri dish). Wash 2 times with water.

Methods for obtaining hair follicle tissue for biochemical analysis

The preparation of hair follicle tissue in milligram quantities is important for biochemical and molecular studies. The first method to epilate hairs with the hair root attached for biochemical studies of wool follicles was first reported by Ellis (1948) who applied a beeswax–resin mixture. The procedure was utilized for investigations of wool follicles and guinea pig hair follicles obtained from freshly removed skin (Rogers, 1959a). The epilation procedure has been improved using self-curing dental wax and is used to obtain gram quantities of wool follicle material from live animals. The domestic sheep is particularly useful as a source because virtually all of the follicles are in the anagen (actively growing) phase and on epilation, the material can be immediately frozen in liquid nitrogen. The only limitation is that the outer root sheath is not removed nor is the hair bulb below about the level of the tip of the dermal papilla. That is, a large proportion of germinative cells remain in the skin. For many purposes this is not important.

Large-scale preparation of follicles from the skin of the sheep

Merino sheep cannot be used because of a tendency to tearing of the skin when the wool is epilated. Fine cross-breeds such as Corriedale or Romney Marsh/Merino are best.

(i)   The wool is shorn down to a length of 2–3 mm using fine animal clippers (Oster, Size 40) on the midside surface and roughly rectangular in shape, measuring 350 mm × 250 mm. Multiple 0.1 ml doses of 2% lignocaine are injected subcutaneously along each edge of the shorn area to minimize any sensitivity of the animal to the ensuing epilation process. Further doses are given if there is any sign of reaction of the animal.

(ii)   Approximately 50 mm × 50 mm areas are marked out with a marker pen.

(iii) The self-curing dental resin (Vertex denture material, Dentimex, Zeist, Holland) is mixed and the marked area of 50 mm × 50 mm is covered with a layer of the paste, pressing the mixture into the stubble of fibers down to the skin surface. A piece of nylon insect screen

---

*Lead citrate (Reynolds): to 10 ml recently boiled water, add 0.2 ml 5N NaOH. Then add 30 mg lead citrate and shake vigorously to dissolve. (Stable for 3 months at 20 °C).

50 mm × 50 mm is pressed into the resin and then this is covered with another layer of resin.

Note: the resin application is carried out in a well-ventilated area (or wearing a mask to absorb fumes) and gloves are worn.

(iv) After the resin is cured (2–5 min) to a rigid state, the patch is grasped, and the skin is slowly pushed away from it with the fingers. The follicles remain attached by the regions of wool fibers embedded in the resin, as they withdraw from the skin. When the patch is totally free from the skin (takes about 30 s), it is immediately plunged into liquid nitrogen. Sometimes, the wool will not epilate without breaking the skin, in which case the procedure is discontinued and another animal chosen. The denuded patches of skin are routinely covered with antiseptic cream (Savlon).

(v) The follicles are stored at −80 °C and, when required, follicle material is readily obtained by scraping them from the hard resin surface using a scalpel blade. The low temperature augments their brittleness, and they can be scraped directly into extraction buffers depending on what is to be prepared. Relatively undegraded RNA or enzymes can be isolated using this procedure.

If it is required, a procedure is available which collects all of the living layers of follicles including the bulb region and outer root sheath. The procedure, devised by Bertolino, Gibbs & Freedberg (1982), is one in which the skin of 12-day old mice is inverted, frozen at liquid nitrogen temperature, and the follicles scraped off from their basal ends together with all of the surrounding dermal tissue. Mice with black fur are used because the intensity of the pigment is a useful guide to the depth of scraping from the dermal side. When the lower follicle regions have been removed, the scraped areas appear black instead of grey. This material was used to isolate follicle mRNA for the preparation of cDNA libraries.

25   Assays for involucrin, transglutaminase and ionophore-inducible envelopes

R. H. RICE

INVOLUCRIN IMMUNOASSAY

Introduction

Involucrin is a major component of the cornified envelope; prior to envelope assembly, it is present as a soluble cytoplasmic protein (Rice & Green, 1979). Polyclonal and monoclonal antibodies to involucrin have been raised by immunizing animals with the purified protein (Rice & Green, 1979, Hudson *et al.*, 1992), and antibodies have also been prepared by injecting synthetic peptides (Simon & Green, 1989; Djian *et al.*, 1993).

For quantitation of involucrin protein, a suitable immunoassay procedure is available which can measure numerous samples in parallel. Using antihuman involucrin antisera prepared as originally described (Rice & Green, 1979), this assay is sensitive to <1 ng of human involucrin. Sensitivity can be particularly important for cells (e.g. malignant keratinocytes) which synthesize low levels of involucrin (Rice, Rong & Chakravarty, 1988; Rubin, Parenteau & Rice, 1989). Alternatively, involucrin concentration can be measured semiquantitatively by immunoblotting, using routine procedures. In favorable cases, the involucrin band can be visualized directly by protein staining if involucrin has been partially purified from crude extracts by solvent treatment or boiling (Etoh, Simon & Green, 1986; Rice & Thacher, 1986; Parenteau, Eckert & Rice, 1987).

Principle

The procedure below has been designed for immunoassay of cultured cell extracts (Parenteau *et al.*, 1987). Typically, cultures are harvested, stored frozen and eventually disrupted by sonication in 1 ml of 10 mM Tris buffer (pH 7.5) – 1 mM EDTA. Aliquots of the soluble extract (or involucrin standards) are incubated with anti-involucrin antiserum. The mixed solution is added to the solid phase antigen, which then is assayed for bound antibody. Thus, as a consequence of competition between soluble and solid phase involucrins, the color yield measures the degree to which involucrin in the sample reduces antibody binding to the solid phase.

In practice, a convenient antibody dilution permits optimal measurement

Keratinocyte methods by Irene Leigh and Fiona Watt
© Cambridge University Press, 1994, pp. 157–165

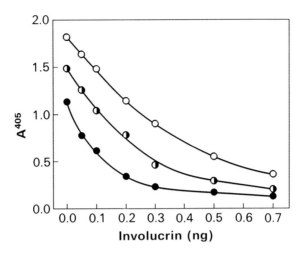

Fig. 25.1. Standard curves for involucrin immunoassay. The color yield (A^{405}) was recorded after incubation of 5 h. The x-axis can also be presented as a log scale, yielding nearly a linear graph of the data. Antiserum dilutions illustrated as 1:1250; 1:2500 and 1:5000.

of involucrin levels in the vicinity of 0.1 ng/well. The best range for measurement depends, in part, upon antiserum concentration since, if more antibodies are present, more involucrin will be required to reduce antibody binding to the solid phase. Fig. 25.1 shows the standard curves obtained with three antiserum dilutions. Less dilution permits measurement of higher amounts of involucrin, while greater dilution gives lower background.

Other species

Antibodies to involucrin are elicited primarily against the repeated 10-amino acid motif that comprises 66% of the human protein (Simon & Green, 1989). These repeated segments differ considerably among mammals (Green & Djian, 1992), enough to reduce dramatically the antigenic cross-reactivity except for very closely related species. Hence immunoassay of involucrin can be futile without antiserum raised to this protein from the species of interest. Primate involucrins, however, retain variable degrees of immunoreactivity toward antihuman involucrin and thus often can be assayed with appropriate reference standards (Parenteau *et al.*, 1987). The shape of the standard curve using antihuman involucrin, reflecting the degree of immunoreactivity, has revealed a high density of determinants in the involucrin of certain species (orangutan and gorilla) and a low density in others, some of which even lack certain epitopes (e.g. gibbon and cebus monkey).

Procedure

Solid phase involucrin

To each well of a 96-well microtiter plate, coated by the manufacturer to maximize protein absorption (flat-bottom Nunc 'Maxisorb' or equivalent), is added 1 ng of human involucrin in 0.1 ml of 0.1 M sodium carbonate buffer (pH 9.6). Chromatographically purified involucrin is used ordinarily, but semipurified fractions perform acceptably if the antiserum employed is sufficiently specific. To each well are then added 10 μl of 0.1 M 1-ethyl-3-(3-dimethylaminopropyl) carbodiimide in water, which promotes adsorption of protein to plastic, and the plate is incubated overnight at 4 °C.

Standard curve

A standard curve is obtained using, for example, involucrin solutions of 0, 0.5, 1.0, 2, 3, 5, 7 and 10 ng/ml in dilution buffer (140 mM sodium chloride – 10 mM sodium phosphate [pH 7.0] – 2 mM EDTA – 1 mM sodium azide – 0.5% Tween 20 – 0.25% gelatin). Aliquots (0.1 ml) of each dilution are added to wells of a second microtiter plate which has not been coated (round bottom Corning low binding or equivalent). Assayed in triplicate, the standard curve occupies 3 rows (A–H). Dilutions of experimental samples to be measured in parallel are arranged similarly. To each well are then added 60 μl of antiserum diluted appropriately in dilution buffer (typically 1:2000). The plate is incubated overnight at 4 °C with gentle agitation.

Color development

After overnight incubation, the solid phase and dilution plates are returned to room temperature. The solid phase plate is emptied by inversion, and to each well is added 0.1 ml of 0.1 M ammonium chloride. After incubation for 30 min, the wells are rinsed 4 times with water and once with dilution buffer. Aliquots (0.1 ml) of involucrin dilutions from the dilution plate are transferred to corresponding locations on the rinsed solid phase plate and incubated 30 minutes at room temperature with gentle agitation. The solution is removed and the solid phase plate is then rinsed 5 times with dilution buffer. To each well is added 0.1 ml of goat antirabbit IgG alkaline phosphatase conjugate. For this purpose, commercial preparations (e.g. Bio-Rad EIA grade or equivalent) typically are diluted 1:5000. The plate is incubated 1 h at room temperature with gentle agitation and then rinsed 4 times with dilution buffer and once with 50 mM sodium carbonate (pH 9.8) – 1 mM magnesium chloride. To each well is then added 0.1 ml of *p*-nitrophenyl phosphate (1 mg/ml) in the carbonate–magnesium buffer.

The development of yellow color, monitored at 405 nm, is usually recorded after 1–5 h using a microtiter plate spectrophotometer.

Acknowledgements

This work was supported by USPHS Grant AR27130 from the National Institute of Arthritis, Musculoskeletal and Skin Diseases.

TRANSGLUTAMINASE ASSAY

Transglutaminases comprise a diverse group of enzymes that cross-link proteins by ε-(γ-glutamyl) lysine isopeptide bonds (reviewed by Greenberg, Birchbichler & Rice, 1991). Since the original finding of transglutaminase activity in mammalian liver (Clarke *et al.*, 1959) and the subsequent discovery that blood clotting factor XIIIa has this activity, the enzymology of transglutaminases has been thoroughly investigated (Lorand, 1972; Folk & Chung, 1973). As used in many laboratories the assay described below is based on the original finding that casein is one of the best acceptor proteins, and putrescine one of the most readily incorporated amines commonly available (Clarke *et al.*, 1959). A number of improvements and refinements have been introduced over the years, including study of synthetic peptide substrates by HPLC with fluorometric detection (cf. Fink, Shao & Kersh, 1992) and immunoassays adaptable to large sample numbers (Michel *et al.*, 1991; Slaughter *et al.* 1992).

Principle

Cell or tissue extracts are incubated with a soluble glutamine-containing substrate protein (α-casein) and a radioactively labeled amine ($[^3H]$–putrescine). After reaction with cell extract in the presence of calcium (a required cofactor) and reducing agent (which protects the active site on cysteine from air oxidation), the labeled protein is separated from unreacted amine by acid precipitation and scintillation counted.

Procedure

Samples of cultured keratinocytes are stored frozen in a nonfrost-free freezer until use. Not only is this convenient, but it also facilitates subsequent homogenization of the cells. Disruption in a Dounce or Ten Broek homogenizer or a sonicator is performed in 20 mM Tris (pH 7.5–8.5)–2 mM EDTA buffer maintained at ice temperature. The crude extract can be assayed directly or after high-speed centrifugation (e.g. 100 000 × g for 45 min) to separate soluble and particulate fractions. Aliquots (typically,

< 50 μg of protein in 25 μl) are assayed in 15 ml conical glass centrifuge tubes in final volumes of 0.25 ml containing 0.5 mg of dimethylcasein, 0.1 M Tris-HCl (pH 8.3), 4 mM CaCl$_2$, 0.4 mM EDTA, 5 mM dithiothreitol, 15 μM putrescine and 0.5 μCi of [^{3}H]putrescine (Rubin & Rice, 1986). After incubation for 30 min at 37 °C, the reactions are stopped by addition of 2.5 ml of 12% trichloroacetic acid – 1 mM putrescine. The samples are chilled on ice (and can be stored in the cold at this stage for several days), pelleted in a clinical centrifuge and the supernatants decanted into radioactive waste (which greatly reduces background values). The pellets are resuspended in 5% trichloroacetic acid – 0.1 mM putrescine, filtered under gentle suction onto Whatman GF/A or equivalent glass fiber filters (presoaked in the wash solution) and washed several times (typically, 5 × 3 ml) with this acid solution. The filters are then rinsed with 3 ml of 95% ethanol, air dried and scintillation counted. Results are commonly expressed in units of nmol of putrescine incorporated per h per mg of extract protein.

Particulate activity

The large majority of activity in cultured keratinocytes is membrane bound (Thacher & Rice, 1985). For best results, this enzyme is solubilized before assay by stirring in buffer containing 0.3–1% nonionic detergent for 1–2 h at 4 °C followed by high-speed centrifugation to remove particulates. (Emulgen 911 has consistently yielded a slightly higher activity (30%) than Triton × -100 or several other nonionic detergents in our hands, but variability among detergent lots has not been tested.) Alternatively, the activity can be solubilized by treatment of the particulate fraction with 1 M hydoxylamine (adjusted to pH 7 with NaOH) at room temperature for 2 h (Chakravarty & Rice, 1989). Removal of particulate material by centrifugation is advisable (although hydroxylamine seriously weakens polycarbonate centrifuge tubes), and the enzyme is quickly desalted by gel filtration through Sephadex G-25 in buffer. Finally, keratinocyte transglutaminase can also be solubilized by mild trypsinization (Thacher & Rice, 1985). Optimal release requires titration of the trypsin solution used, but generally treatment for 15 min at 20–37 °C with TPCK–trypsin (0.1–1 μg in 0.2 ml final volume) works well with the pariculates from one confluent 10 cm culture of human epidermal cells (Chakravarty, Rong & Rice, 1990). Whichever solubilization method is employed, the remaining activity in the particulate fraction often is assayed to determine efficacy of the treatment. While nonionic detergent extraction would be expected to be effective in general for membrane proteins, success of the hydroxylamine and trypsin treatments depends upon specific properties of keratinocyte transglutaminase (thioesterified fatty acid and proteolytic hypersensitive site(s) near the amino terminus, respectively) and thus may not be applicable to novel membrane-bound isozymes.

Acceptor substrates

Almost any protein is anticipated to participate in transglutaminase cross-linking through its e-amino residues that can serve as amine donors. Relatively few proteins provide glutamine acceptor residues, however. The primary factor in determining whether a given peptide-bound glutamine serves as an amine acceptor appears to be conformational availability, but nearby charged residues can influence greatly its reactivity in the presence of a given transglutaminase (Coussons *et al.*, 1992). α-Casein is a good acceptor substrate for known transglutaminases in mammalian keratinocytes, but it is not guaranteed to work well for novel isozymes. For example, it serves poorly in this role in assays of activity derived from nematodes (Mehta *et al.*, 1992) and plants (Signorini, Beninati & Bergamini, 1991). Transglutaminases show marked substrate preferences. Thus, β-casein and certain glutamine-containing peptides derived from it are good acceptor substrates for plasma factor XIIIa but are unreactive toward tissue transglutaminase (Gorman & Folk, 1980). Conversely, synthetic peptides based on a fibronectin cross-linking site selectively inhibit tissue translutaminase in the presence of factor XIIIa (Parameswaran *et al.*, 1990).

Acceptor substrates

The usual assays exploit the ability of transglutaminases to employ primary alkylamines as nucleophiles to attack the enzyme glutamate thioester intermediate. Putrescine (1,4-diamino-n-butane) is commonly used because it is available in high specific activity and is bifunctional. Indeed, transglutaminases have been shown to incorporate this and other polyamines into cross-linked structures *in vivo* (Piacentini *et al.*, 1988). A variety of small amines have been shown to be reactive in assays, of which some, such as histamine (Fesus *et al.*, 1985), have been observed to be incorporated in intact cells as well. Since amine incorporation can occur through other reactions besides transglutaminase-mediated cross-linking (e.g. as a consequence of amine oxidation), direct identification of the cross-linked product is advisable in novel instances (Lorand & Conrad, 1984) and in many cases has been accomplished. Fluorescent amines, chiefly dansylcadaverine, are often used in place of radioactive ones for fluorimetric detection and can even provide epitopes for immunochemical detection of transglutaminase substrates (Kvedar *et al.*, 1992).

A consequence of the prevalence and availability of ε-amino groups of proteins is that the measured activity (radioactive putrescine incorporation, for example) of crude extracts is greatly reduced at high protein concentration. Thus, small rather than large aliquots are assayed whenever possible, and it is advisable to demonstrate linearity of the assay with increasing amounts of extract to be assured of the meaningfulness of a measured specific activity. The ε-amino groups in casein also compete with the intended amine donor, a difficulty which is minimized by using dimethyl-casein. Briefly, α-casein is subjected to reductive methylation (Means & Feeney, 1968) by treating it (150 mg in 10 ml of 0.1 M sodium borate, pH 9) with formaldehyde (0.6 ml of a 1.4 M solution) at 0 °C in the presence of sodium borohydride (0.3 ml of a 50 mg/ml solution) with constant stirring in the presence of a trace of n-octanol to retard foaming. After four cycles of borohydride/formaldehyde addition at 5 min intervals, yielding 80% amino group modification, the protein is dialyzed extensively and stored frozen in aliquots until use. This modification improves the sensitivity of the assay 5-fold (Rice *et al.*, 1988; Klein, Guzman & Kuehn, 1992).

IONOPHORE-INDUCIBLE
ENVELOPE ASSAY

A characteristic feature of keratinocyte differentiation (cf. Green, 1979), cross-linked envelopes are synthesized ordinarily by a small fraction of cultured keratinocytes (Sun & Green, 1976). However, when the cells are held in suspension (Green, 1977) or treated with inhibitors of protein synthesis (Rice & Green, 1978), a majority form envelopes. This process can be accelerated in surface culture by addition of ionophores such as A23187, X537A, gramicidin S (Rice & Green, 1979) and even melittin (Rice *et al.*, 1988). Envelope formation in culture usually is attributable to activation of the membrane-bound keratinocyte transglutaminase (Thacher, Coe & Rice, 1985), but tissue transglutaminase also appears capable of forming similar structures visible microscopically (Levitt *et al.*, 1990). *In vivo*, the soluble epidermal transglutaminase may also contribute to envelope formation or stability (Kim *et al.*, 1990).

While transglutaminase activity is essential, the substrate proteins required for envelope formation are not well defined. This reflects in large part the many proteins observed to be incorporated into these structures (cf., Reichert, Michel & Schmidt, 1993). In addition, envelopes may actually comprise a number of distinct entities. For example, two morpho-logical types are distinguishable from epidermis (Michel *et al.*, 1988), and analogous but distinctive structures are visible in nail and hair. Envelope composition may also reflect different physiological states of the cell (Michel *et al.*, 1987; Nagae *et al.*, 1987). Since conditions in culture only approximate those *in vivo*, and accelerated envelope formation in any case may perturb the normal process (Warhol *et al.*, 1985; Yuspa *et al.*, 1989), the assay described below is only a general indicator of the state of cell differentiation. Measurement of specific protein markers often is preferable when possible. However, ionophore-inducible envelope formation measures the participation of a number of possible components at once, which can be advantageous in some circumstances.

Principle

Cross-linked envelopes form spontaneously in cultured normal epidermal keratinocytes at a low level, but most of the cells are competent to synthesize them. To measure the extent to which this stage of differentiation is reached, the cells are treated with an ionphore which raises the internal calcium concentration and thereby results in transglutaminase activation. After several hours, the cells are then treated with an ionic detergent and reducing agent, which dissolve the keratins and other noncross-linked proteins. Resistant cross-linked structures are then visible microscopically and are scored.

Procedure

For qualitative purposes using surface cultures, the simplest approach is to rinse the cells with serum free medium and add serum free medium containing X537A (50 μg/ml). A stock solution of the ionophore (5 mg/ml) can be stored in ethanol for months at 4 °C. A more specific calcium ionophore, A23187 is equally effective but is stored in dimethylsulfoxide since it is less soluble in ethanol (and aqueous media). After 5 h in the culture incubator the medium is adjusted to 2% in SDS and 20 mM in dithiothreitol. After 20 min or more at room temperature, during which time the opaque cell interiors dissolve, envelopes can be visualized microscopically. A semiquantitative estimate of the envelope content can be obtained spectrophotometrically. This is useful when the original cultures are highly confluent, so that trypsin disaggregation prior to ionophore treatment is difficult (Rice & Cline, 1984). The treated cells and medium are first passed through a fine needle (25 gauge), shearing the DNA and reducing the viscosity. Envelopes are pelleted by high speed clinical centrifugation and rinsed in 0.1% SDS to remove small particulates. The envelopes are resuspended in a known volume of 0.1% SDS and light scattering is measured by A^{340}. The readings slowly decline due to gradual settling of the envelopes.

For more quantitative purposes, cultures are trypsinized. The disaggregated cells are rinsed in serum-containing medium and resuspended at 10^6/ml in serum free medium containing X537A (50 μg/ml). To obtain an accurate measurement of the cell concentration, an aliquot is counted quickly in a hemacytometer before clumping occurs. The cell suspension is incubated for 2 h in the culture incubator and then adjusted to 2% in SDS and 20 mM in dithiothreitol. After 20 min or more, the envelopes are counted in the hemacytometer by phase contrast or Nomarski optics. A source of inaccuracy at this stage, particularly when a small fraction of the cells are envelope-competent, is a tendency for the envelopes to become clumped with DNA, making it difficult to count a representative aliquot. The clumping can be minimized by vortexing the sample or otherwise shearing the DNA before counting. In addition, the buoyant envelopes often float slowly across the field, so one must count them in a systematic fashion so as to compensate for this motion. With some practice, reproducible counts are obtainable.

Acknowledgements

This work was supported by USPHS Grant AR27130 from the National Institute of Arthritis, Musculoskeletal and Skin Diseases.

26 Production of recombinant human involucrin using the glutathione-S-transferase fusion system

N. SHEIBANI and B. L. ALLEN-HOFFMANN

Introduction

Native involucrin protein can be purified to homogeneity from skin or cultured keratinocytes (Rice & Green, 1979). However, soluble involucrin is highly susceptible to proteolytic degradation and the purification protocol is lengthy. For these reasons, we used the pGEX-2T prokaryotic expression system to generate recombinant involucrin protein. The expression vector contains the C-terminus of the glutathione-S-transferase (GST) gene from *Schistosoma japonicum* which provides an affinity tail on the involucrin fusion polypeptide (Smith & Johnson, 1988). The GST is expressed from a *tac* promoter, allowing inducible expression of the fusion protein. The vector also expresses the *lac* I^g gene which suppresses *tac* promoter activity. Thus, the expression vector may be used in any *E. coli* strain. The vector also contains a sequence immediately downstream of the GST and upstream of the multiple cloning site which encodes a specific amino acid sequence susceptible to thrombin cleavage. Following induction of expression with isopropyl-β-D-thiogalactopyranoside (IPTG), the involucrin fusion protein is easily recovered from bacterial lysate by affinity chromatography on immobilized glutathione. Pure involucrin is obtained by site-specific thrombin cleavage of the polypeptide from the GST tail within the affinity matrix.

The pGEX system can readily and simply be used for one-step purification of a large quantity of involucrin under nondenaturing conditions. The recombinant involucrin protein is indistinguishable from native involucrin as a solid phase or as a standard for competitive immunosorbent assays. In addition, we found this system advantageous for the production of high titer polyclonal antibodies against involucrin in rabbits.

Expression vector

The construction of the expression vector is outlined in Fig. 26.1. The plasmid p lambdaI-3 was obtained from ATCC (Rockville, MD). This plasmid contains a 6 Kbp HindIII/BamHI genomic fragment that contains the sequence for the human involucrin gene (Eckert & Green, 1986). The

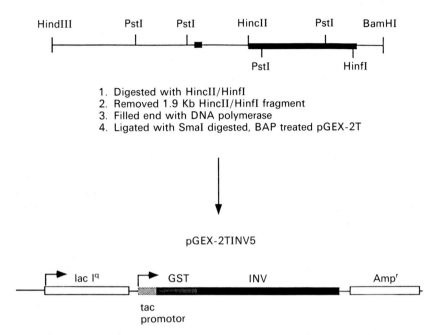

Fig. 26.1 Construction of the human involucrin expression vector.

entire involucrin coding region was removed by digesting this fragment with HincII/HinfI. The 1.9 Kbp HincII/HinfI fragment was blunt ended by DNA polymerase Klenow fragment and ligated in the correct reading frame and orientation to SmaI digested, dephosphorylated pGEX-2T plasmid DNA. The nucleotide sequence was confirmed by dideoxynucleotide sequencing of the recombinant plasmid DNA. A recombinant plasmid pGEX-2TINV5 was obtained and used for expression of the involucrin fusion protein.

Expression and purification of involucrin

A modification of the methods of Smith and Johnson (1988) and Abath and Simpson (1991) was used to express and purify involucrin from *E. coli*. The recombinant or control plasmids were transformed into *E. coli* MAX efficiency DH5aF'IQ competent cells (Gibco BRL, Grand Island, NY) which overproduce *lac* I^g. Overnight culture of *E. coli* transformed with pGEX-2T or pGEX-2TINV5 in Luria Broth (LB) were diluted 1:10 in 20 ml of LB and grown for 1 h at 37 °C before adding IPTG to 1 nN. After an additional 4 h of growth, the cells were pelleted, resuspended in 0.5 ml of PBS (150 mM NaCl, 16 mM Na_2HPO_4, 4 mM NaH_2PO_4, pH 7.3) and transferred to a microfuge tube. Cells were lysed by freezing/thawing in a dry ice/ethanol bath (5 times, 2 min cycles). The cell lysate underwent two cycles of sonification in a Heat System XL2020 sonifer at a setting of 7 for 2 min each cycle. Triton X-100 (10% w/v) was added to 1%, the solution was mixed and centrifuged at 16 000 × g for 5 min at 4 °C. The supernatant was transferred to a clean tube and mixed at room temperature on a platform rocker with 0.3–0.5 ml of 50% glutathione-agarose beads (Sigma, G-4510). Following absorption for 5–10 min, the beads were collected by a brief centrifugation in a microcentrifuge (10 s 16 000 × g), and washed 5 times with 1 ml of PBS to remove contaminating proteins.

To remove involucrin from the GST tail, the beads from above were washed once with 1 ml of thrombin buffer (50 mM Tris-HCl, pH 8.0, 150 mM NaCl, 2.5 mM $CaCl_2$) and resuspended in thrombin buffer to 80% beads and incubated with 10 μl (6 NIH units) of thrombin (Sigma, T-3010) at room temperature for 1 h with continuous mixing. The beads were pelleted by centrifugation (10 s, 16 000 × g) and the supernatant containing the thrombin-cleaved involucrin was saved. The beads were washed an additional four times with 0.1 ml of 50 mM Tris-HCl, pH 8.0, 5 mM EDTA and all washes were pooled with the supernatant from the initial thrombin cleavage step. The presence of EDTA effectively inhibits the activity of the calcium-dependent thrombin enzyme. To remove thrombin, the pooled washes were placed in a boiling water bath for 5 min and subsequently centrifuged (5 min, 16 000 × g) at 4 °C. The protein concentration was determined by the method of Lowry (Lowry *et al.*, 1951). Proteins were also analyzed by sodium dodecyl sulfate–9% polyacrylamide gels (SDS–PAGE) in the absence of reducing agent and visualized by Coomassie blue staining and silver staining.

Antibody production

Antibodies against human involucrin were raised in New Zealand white rabbits. Two sets of rabbits were injected subcutaneously at multiple sites on the back with either 0.6 ml of agarose-bound involucrin fusion protein complex (from 20 ml of culture) or 1 ml of purified involucrin protein (from 40 ml of culture). The immunogens were mixed 1:1 with complete Freund's adjuvant. These amounts corresponded to approximately 100 μg of the agarose-bound involucrin fusion protein in a 50% bead mix and 200 μg of the thrombin-released purified involucrin. The animals were given boost injection 4 weeks later with approximately the same amount of antigen, mixed 1:1 with incomplete Freund's adjuvant. A week later the animals were bled and the serum was prepared as described by Harlow and Lane (1988).

Enzyme-linked immunosorbent assay for involucrin

An enzyme-linked immunosorbent assay originally described by Parenteau *et al.* (1987), and modified by Sheibani, Rhim & Allen-Hoffmann (1991), was used to test both the utility of recombinant involucrin as a solid phase and as a standard for immunoassay. The presence of anti-involucrin antibody in the serum collected from immunized rabbits was determined using this assay as well. Polyclonal rabbit antihuman involucrin antibody and purified human involucrin standard were obtained as part of an involucrin ELISA assay kit (BT-650) from Biomedical Technologies, Inc (Stoughton, MA). Goat antirabbit IgG covalently coupled with alkaline phosphatase (Sigma A-8025) was substituted for the kit alkaline phosphatase in all assays. The appropriate dilution of alkaline phosphatase is determined for each lot (usually a 1:2000 dilution). A_{405} was read after development of sufficient color (< 2 h).

Results

Expression of recombinant human involucrin

The involucrin fusion protein was observed following induction of the *tac* promoter with IPTG. Expression was optimal within 4–6 h after addition of IPTG to 1 mM. Polyacrylamide gel analysis of the cell lysates revealed a novel 165 kDa protein, consistent with the expected size of the recombinant protein (Fig. 26.2, lane 1). To purify recombinant involucrin, the cleared lysate was incubated with glutathione–agarose beads, and pure involucrin protein was obtained by thrombin cleavage. Recombinant involucrin migrated with an apparent size of a 140 kDa protein (Fig. 26.2, lane 2). The 27 kDa band is the glutathione-S-transferase fragment generated after thrombin digestion and the 36 kDa band is thrombin. Silver staining of recombinant involucrin demonstrated that there were no degradation products of involucrin generated by the thrombin cleavage step. Neither extended incubation time of the involucrin fusion protein with thrombin (24 h) nor failure to add EDTA to inhibit thrombin activity resulted in involucrin degradation product (Fig. 26.2, lanes 3 and 4). These results were expected in that involucrin does not contain either the anion exosite or the cleavage-recognition sequences for this calcium-dependent, site-specific protease. Since involucrin is stable and soluble at high temperatures, boiling the thrombin-cleaved involucrin preparation followed by centrifugation, effectively removes this enzyme (Fig. 26.2, lane 5). We obtained 50–100 μg of involucrin from 20 ml of bacteria culture.

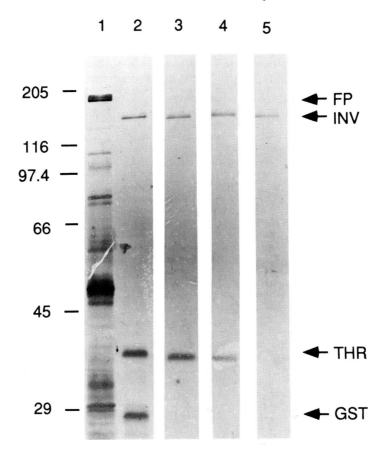

Fig. 26.2.   Purification and stability of human involucrin expressed as a GST fusion protein. Cells transformed with pGEX-2TINV5 were grown and involucrin fusion protein was purified. Samples (10 μl) were analyzed by a non-reduced 9%-SDS–PAGE gel and visualized by silver staining. **Lanes 1**, supernatant from total cell extract; **2**. recombinant involucrin after 1 h incubation of matrix-bound involucrin with thrombin; **3**. thrombin-released involucrin in pooled washes after 24 h incubation at rt without EDTA; **4**. thrombin-released involucrin in pooled washes after 24 h incubation at rt with EDTA; **5**. thrombin-released involucrin in pooled washes boiled for 5 min then incubated without EDTA at rt for 24 h. The position and size (kDa) of M_r markers are indicated.

Use of recombinant involucrin in immunoassay and for antibody production

In a competitive immunosorbent assay for human involucrin, both native and recombinant involucrin were used to generate standard curves (Fig. 26.3). Recombinant and native involucrin solutions of known concentration

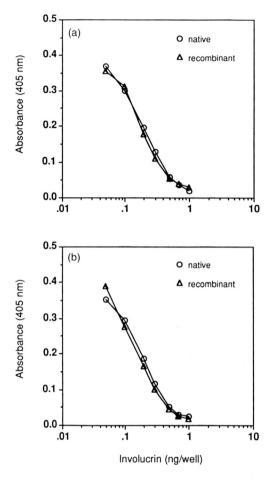

Fig. 26.3. Use of recombinant involucrin as solid phase and as standard in competitive immunosorbent assay. Anti-involucrin antibody was from rabbits immunized with matrix-bound involucrin. (*a*) Native involucrin solid phase; (*b*) Recombinant involucrin solid phase. Standard curves generated with native or recombinant involucrin solutions.

generated identical standard curves when native involucrin isolated from cultured human keratinocytes (Fig. 26.3(*a*) or recombinant involucrin (Fig. 26.3(*b*) was used as the solid phase. Therefore, recombinant human involucrin can be substituted for native involucrin in immunoassays.

Purified peptides (Johnson *et al.*, 1989) and the purified GST-fusion proteins (Toye *et al.*, 1990) generated by this system have been successfully used by others for immunization and antibody production. We took advantage of a step in the recombinant involucrin purification protocol that employs affinity chromatography using glutathione–agarose beads.

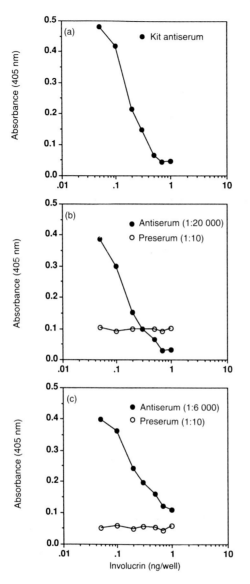

Fig. 26.4. Screening of antisera from rabbits immunized with recombinant involucrin. A week after the first boost, blood was collected from each rabbit and different dilutions of the antisera in TBS (0.15 M NaCl, 0.01 M Tris-HCl, pH 7.4 with 0.1% BSA) were used as the source of antibody to generate standard curves in a competitive immunosorbent assay. (*a*) Commercial antibody; (*b*) Antiserum from a rabbit immunized with glutathione-agarose bound involucrin fusion protein (1:20 000 dilution); (*c*) Antiserum from a rabbit immunized with thrombin-released recombinant involucrin protein (1:6000 dilution). The pre-immune sera were used at a 1:10 dilution.

Cross-linking of proteins to agarose beads has been used to enhance antigenicity of some proteins (Harlow & Lane, 1988). Since the GST tail is small relative to involucrin, we predicted that the fusion protein, as a complex with agarose beads, would be at a distance from the matrix. This would enhance the probability that the involucrin would be recognized in various surface configurations and elicit a broad antigenic response. Our approach also circumvents use of thrombin to cleave the recombinant protein, thus eliminating a step in protein purification. We consistently found that the glutathione–agarose bound involucrin fusion protein elicited an apparently stronger immune response when compared to the thrombin-released involucrin protein even though less matrix-bound involucrin was used for immunization and antibody production (Fig. 26.4). Recent studies conducted with mice support our findings that matrix-bound recombinant proteins are superior immunogens (Oettinger, Pasqualini & Bernfield, 1992). We routinely use antiserum from rabbits immunized with matrix-bound involucrin at dilutions of 1:16 000 for the immunosorbent assay. Therefore, the affinity matrix can be exploited to minimize the number of protein purification steps and to enhance the antigenicity of the involucrin polypeptide.

Acknowledgements

We thank Sandy Schlosser for her technical assistance and Samir Kharebandeh for his help during early stages of this project. We thank Drs Alicia Domberoski and Jeff Jones for their help in sequencing the recombinant plasmid DNAs. This work was supported by Grant R29 AR40284 (to B.L. A-H); N.S. is a trainee on Grant HD07118 from NIH.

Part V

INTRODUCTION OF FOREIGN GENES INTO KERATINOCYTES

Introduction

Expression of exogenous genes in keratinocytes is a powerful approach for studying the functions of those genes and their effects on keratinocyte behavior. In addition, it has potential therapeutic applications in gene therapy since cultured keratinocytes can be used as grafts (see Part 1). Expression vectors containing the cDNA of interest are introduced into keratinocytes and the cells are examined within a few hours or days (transient assays) or after prolonged selection for cells that are stably expressing the gene product. The major difficulty encountered with human keratinocytes is the low efficiency with which the vectors are taken up and expressed. The most common technique, coprecipitation of DNA with calcium phosphate, yields satisfactory results if the introduced genes confer a selective growth advantage (for example, the transforming genes of HPV 16; Pei, Gorman & Watt, 1991), or if the target cells are an established keratinocyte line (see, for example, Albers & Fuchs 1989). An alternative approach is to use retroviral vectors, which can infect keratinocytes with high efficiency (Garlick *et al.*, 1991, Gerrard *et al.*, 1993). In this section, a protocol for retroviral infection is given (see Garlick, Chapter 29), together with lipid- or polybrene-based transfection for stable (see Sexton & Bouvard, Chapter 28) or transient (see Carroll, Chapter 27) gene expression.

27 Keratinocyte lipidfection for transient transfections

J. CARROLL

This protocol is adapted for use on keratinocytes from the original method of Rose, Buonocore & Whitt (1991) for transfecting various epithelial cells. Approximately 5–10% transfection efficiency is achieved in primary keratinocytes, some squamous cell carcinoma (SCC) lines, and a variety of other cell types (such as primary fibroblasts, HeLa, and MCF-7 cells). This protocol is at least two-fold more efficient than current transfection methods involving polybrene and calcium phosphate. Additionally, the conditions seem less harsh than these protocols in that no 'shock' is used and the morphological appearance of the cells after transfection is better. This method is, I believe, preferable to calcium phosphate for transient transfections since any effects of high levels of calcium on gene expression are avoided. Furthermore, the reagents used are much less expensive than the 'lipofectin' system (Gibco-BRL) and give comparable transfection efficiency.

Preparation of lipids

The two lipids used are PtdEtn (dioleoyl-L-α-phosphatidylethanolamine) and DDAB (di-methyldioctadecylammonium bromide). Both of these were purchased from Sigma; DDAB as a powder, and PtdEtn as a chloroform concentrate.

(i)   10 mg DDAB is dissolved in 2.5 ml PtdEtn-chloroform solution (supplied as 10 mg/ml in chloroform = 25 mg) in a *glass* tube.

(ii) The solution is then evaporated to dryness using a Speed-Vac (Savant) or vacuum aspirator (this is a chloroform solution and you must check that it does not represent an explosion hazard in the vacuum apparatus). The evaporation time of this step is crucial. Evaporate just to the point where all the chloroform is gone (2 h in a Speed-Vac for 2.5 ml is usually sufficient). *Do not* evaporate overnight because overdrying renders the lipids much more resistant to dissolving in the next step.

(iii) Resuspend the evaporated lipids in 20 ml sterile distilled water (in a sterile 50 ml conical tube) and sonicate on *ice* using a cell sonicator (Virsonic Cell Disruptor 16-850, 50 V setting using a microprobe).

Sonicate for a few minutes until the clumps break up and the solution is homogeneous and full of translucent swirls. (*Do not* sonicate much longer than is necessary).

Cell transfection

(i)   Seed keratinocytes or SCC cells at approximately 5×10^5 cells/per 60 mm diameter petri dish (p60) on a layer of irradiated 3T3 feeder cells. Transfect cells at 20–60% confluence (usually 2–3 days post-seeding). Feed dishes the night before transfection.

(ii) For a single p60 (for a p100, multiply everything $\times$ 2.5), add 45 μl lipid mix to 1.5 ml serum-free medium (this may contain any additives usually used for keratinocyte growth except serum) in a tube. Mix gently. Add 12 μg DNA (can use from 8–20 μg; more DNA results in more expression). Mix gently. Incubate at room temperature for 5–10 min.

(iii) Wash cells once in serum-free medium. Add DNA/lipid mix to cells. Incubate at 37 °C for 3–4 h; rock plate every hour or so to ensure even coverage. After incubation, add 1.5 ml (i.e. an equal volume) of growth medium containing serum. Incubate for 2–3 more h, then aspirate medium and wash twice with growth medium. Add 4 ml of growth medium and incubate at 38 °C.

(iv) Cells can be harvested 48–72 h later for transient assays. The cells should be confluent at this point.

Note

If the transfection efficiency is lower than 5%, make up another batch of lipids. For unknown reasons, occasionally some batches do not work well. It is a good idea to titer your cells for DDAB toxicity, and use the highest concentration of DDAB that does not adversely affect the cells. The quantity reported here represents the optimal DDAB concentration for normal human keratinocytes. For prolonged storage, lipids are kept under nitrogen gas at 4 °C.

28   Stable transfection of human keratinocytes: HPV immortalization

C. SEXTON and V. BOUVARD

Lipofection

(i)   Plate cells 48 h prior to transfection at 2×10^5 cells per 100 mm dish or 25 cm^2 flask (T25) in KCM (complete keratinocyte culture medium; see Navsaria *et al.*, Chapter 1) on a 3T3 feeder layer. This should result in small colonies of 10–15 cells and about 20–30% confluency on the following day. Seeding results are, however, dependent upon the degree of stratification of the original culture which should not be overconfluent when passaged.

(ii)   One day after plating, remove feeders by treating with versene for 3–5 min. Replace medium with either KGM (serum-free keratinocyte growth medium/Clonetics, see Mitra & Nickoloff, Chapter 4). Stratification may inhibit uptake of transfected DNA and the use of KGM, which has a low concentration of calcium ions, decreases the degree of stratification and increases transfection efficiency.

(iii)   4 h prior to adding the transfection mixture replace the medium with 2 ml of Optimem (Gibco).

(iv)   For transformation studies, we routinely use 10 μg of plasmid DNA. If a selectable marker plasmid is to be included, 10 μg of that is also used. The transfection mixture is made as follows:
- Solution A: 1.5 ml Optimem + X μg plasmid DNA.
- Solution B: 0.5 ml Optimem + 30 μl Lipofectin Reagent (BRL).
- Solution C: Add soln. B to soln. A = 2 ml.

(v)   Replace medium on the cells with solution C and incubate for 12–24 h at 37 °C. Wash cells twice with serum free DMEM and replace with RM+ and irradiated 3T3 feeders.

(vi)   Solutions should be made fresh immediately prior to transfection. If multiple transfections are being performed solutions can be scaled up and aliquoted into appropriate tubes.

(vii)   Selection can be initiated 3–4 days posttransfection. We routinely use G418 selection at 0.1 mg/ml on human oral keratinocytes.

Keratinocyte methods by Irene Leigh and Fiona Watt
© Cambridge University Press, 1994, pp. 179–180

Polybrene

The main advantage of this procedure over lipofection is the relatively inexpensive materials required. The polybrene-DMSO transfection involves two steps: first polybrene (1,5-dimethyl-1,5-diazaundecamethylene polybromide) which is a polycation absorbs the polyanionic DNA molecules to the cell surface; DMSO treatment then enhances uptake of the absorbed DNA by increasing the permeability of the cell membrane.

(i)   Plate keratinocytes the previous day at 8×10^5 per 100 mm dish in the absence of feeders.
(ii) Wash with PBS.
(iii) Add 10 μg of DNA to 4 ml DMEM containing 30 μg/ml polybrene.
(iv) Place the mixture directly onto the cells and incubate 6 h at 37 °C.
(v)   Remove the transfection medium and replace with 5 ml DMEM + 10% FCS +30% DMSO for 4 min.
(vi) Rinse plates 2 × with 10 ml keratinocyte growth medium. Add back feeders and growth medium and culture normally.

Polybrene stock solution
30 mg/ml (Sigma) in water. Store for up to 2 weeks at 4 °C.

29   Retroviral vectors

J. GARLICK

Introduction

Two useful reviews about the preparation and uses of retroviral vectors
and packaging lines have been written by Price (1987) and Morgenstern
and Land (1991). For examples of specific applications of retroviral vectors
for infection of keratinocytes, see Garlick *et al.* (1991) and Gerrard *et al.*
(1993). When working with retroviruses that have a human host range, it
is important that stringent safety precautions are taken to avoid transmission
of the viruses to laboratory personnel.

The efficiency of keratinocyte transduction by retrovirus vectors is a
function of the keratinocyte growth fraction and the titer of the infecting
retroviral vector. Since retroviruses will only infect cells that are actively
dividing it is important that a maximal number of cells be dividing at the
time of infection. The optimal time of infection therefore needs to be
determined for different cell types and strains. The number of cells plated
in the following protocol was determined as optimal for early passage
human foreskin keratinocytes. The number of cells that should be seeded
for infection will be a function of their plating efficiency and growth rate.

Cell stocks

RETROVIRAL PRODUCER CELL LINE: Carrying retrovector and gene of
interest.

NEOMYCIN RESISTANT FEEDER CELL (Neo^r 3T3): A stable line of 3T3 cells
containing the gene for neomycin phosphotransferase that confers neomycin resistance (Neo^r). These cells are generated by transfecting 3T3 with
a Neo^r plasmid and selecting with G418 sulfate antibiotic (Geneticin).

KERATINOCYTES: Should be early passage cells.

Day 1

(i)   Irradiate Neo^r 3T3 cells and plate at 2×10^6 in 100 mm dish(P100)
      in complete keratinocyte medium (Wu *et al.*, 1982).
(ii) Plate retroviral producer cells at 2×10^6 in 3T3 media (DME + 10%
      fetal calf serum + pen/strep).

Keratinocyte methods by Irene Leigh and Fiona Watt
© Cambridge University Press, 1994, pp. 181–183

Day 2

(i) Keratinocytes are trypsinized, suspended as single cells and plated onto Neo^r 3T3. The keratinocyte seeding densities depend on the viral titer of the infecting virus and whether sheets or individual colonies of cells (for cloning) are desired.

- For viral titers of 10^7 cfu/ml
  use 3000 keratinocytes for sheets/p100
  600 keratinocytes for colonies /p100
- For titer of 10^6 cfu/ml
  use 30 000 keratinocytes for sheets/p100
  6000 keratinocytes for colonies/p100
- For titer of 10^5 cfu/ml
  use 300 000 keratinocytes for sheets/p100
  60 000 keratinocytes for colonies/p100

(ii) Change medium on producer line several hours before infection by adding a minimal amount of media (6 ml/p100) to producer cells. Producer cells should be used as confluent as possible to ensure maximal titers.

Day 3

KERATINOCYTE INFECTION: Remove medium from keratinocytes and add 2 ml of supernatant from producer cells that has been filtered through a 0.45 μm filter and that has 1.6 μl polybrene (10 mg/ml stock) added to it. Gently rock plates every 15 min for 2 h while incubating at 37 °C. After this, remove the viral supernatant and replace it with 10 ml complete keratinocyte medium.

Day 6 to Day 9

Keratinocytes must be allowed to go through two cell doublings before selection; this is the time required for expression of the transduced gene. Select by adding G418-containing medium with a concentration of 800 μg/ml active G418.

Day 19 to Day 22

Colonies will grow out in selective medium and can usually be cloned on day 19–22. Cells can be fed with G418 for 10 days (minimum of 3 feedings) and afterwards the G418 can be replaced with pen/strep.

Notes on retroviral transduction

Keratinocyte transduction rates have been shown to vary linearly with viral titer. Attempts at concentrating virus have proven to be difficult, yet there are reports of increases in titers as great as two orders of magnitude. Retroviral vectors are engineered to include *cis*-acting sequences known as the ψ-packaging sequence which are required for efficient virus encapsidation. Vectors containing an extended ψ-packaging sequence (Miller & Rosman 1989) demonstrate higher titers than those that do not contain the extended ψ-packaging sequence. Finally, the size of the gene of interest inserted into the vector will also determine the titer of the infectious virus.

The above protocol describes the stable transduction and selection of keratinocytes with a Neo^r-containing transgene. It is important that keratino-cyte colonies are sufficiently small at the time of selection. Selection with G418 usually requires several days before cell death occurs and cells will continue to grow during that time. As a result, colonies that have become confluent tend to pull off the plate together and this may result in the loss of transduced colonies as well. As an alternative to the above selection technique, keratinocytes can be infected on normal 3T3 feeders and subsequently passaged on DAY 6-9 directly onto Neo^r feeders in G418 media. If selection is not desired, cells can be maintained in complete keratinocyte medium.

In addition to the above protocol for transduction of keratinocytes with retroviral vector supernatant, it is possible to infect cells by coculturing them with gamma-irradiated or mitomycin C treated viral producer cells used as a feeder layer. Transduction rates are generally higher using this technique. However, plating efficiencies are somewhat lower on the viral producers than on Neo^r 3T3 feeders.

Retroviral vector packaging cell lines constituitively express the retroviral genes necessary for virus assembly and replication that have been deleted from the vector. The cell lines are commercially available and can be grown under conditions similar to those under which most fibroblast cell lines are maintained. These lines generate retrovirus with either a limited host-range (ecotropic) or a broad host-range (amphotropic) for infection. The packaging cell line to be used will therefore be determined by the species of the cell to be transduced. Cationic liposomes have been used to mediate retroviral infection to cells that are resistant to retroviral infection. Using this system, the host range of infectious virus can be extended but this is at the expense of viral titers which are lowered considerably.

Genes transferred by retrovirus-mediated transduction are expressed in both differentiated and undifferentiated keratinocytes and expression of these genes appears to be stable in culture with serial passage. In a small percentage of colonies, a mosaic pattern of transgene expression has been observed where only a portion of cells in a colony express the gene (Garlick *et al.*, 1991). This phenomenon may be due to promoter shutoff and it can possibly be ameliorated by the use of endogenous promoters or vectors utilizing single transcriptional units.

References

Aarden, L.A., De Groot, E.R., Schaap, O.L. & Lansdorp, P.M. (1987). Production of hybridoma growth factor by human monocytes. *European Journal of Immunology*, **17**, 1411–16.

Abath, F.G.C. & Simpson, A.J.G. (1991). A simple method for the recovery of purified recombinant peptides cleaved from glutathione-S-transferase-fusion proteins. *BioTechniques*, **10**, 178.

Adams, J.C. & Watt, F.M. (1989). Fibronectin inhibits the terminal differentiation of human keratinocytes. *Nature*, **340**, 307–9.

Adams, J.C. & Watt, F.M. (1990). Changes in keratinocyte adhesion during terminal differentiation: reduction in fibronectin binding precedes $\alpha_5\beta_1$ integrin loss from the cell surface. *Cell*, **63**, 425–35.

Adams, J.C. & Watt, F.M. (1991). Expression of β_1, β_3, β_4 and β_5 integrins by human epidermal keratinocytes and non-differentiating keratinocytes. *Journal of Cell Biology*, **115**, 829–41.

Akhurst, R. (1993). Localisation of growth factor mRNAs in tissue sections by *in-situ* hybridisation. In *Growth Factors: A Practical Approach*, ed. I. A. McKay & I.M. Leigh, Practical Approach Series. Oxford: Oxford University Press.

Al-Bawari, S.E. & Potten, C.S. (1976). Regeneration and dose-response characteristics of irradiated mouse dorsal epidermal cells. *International Journal of Radiation Biology*, **30**, 201–16.

Albers, K. & Fuchs, E. (1989). Expression of mutant keratin cDNAs in epithelial cells reveals possible mechanisms for initiation and assembly of intermediate filaments. *Journal of Cell Biology*, **108**, 1477–93.

Alitalo, K., Kuismanen, E., Myllylä, R., Kiistala, U., Asko-Seljavaara, S., & Vaheri, A. (1982). Extracellular matrix proteins of human epidermal keratinocytes and feeder 3T3 cells. *Journal of Cell Biology*, **94**, 497–505.

Allen-Hoffmann, B.L. & Rheinwald, J.G. (1984). Polycyclic aromatic hydrocarbon mutagenesis of human epidermal keratinocytes in culture. *Proceedings of the National Academy of Sciences, USA*, **81**, 7802–6.

Angerer, L.M., Cox, K.H. & Angerer, R.C. (1987). Demonstration of tissue-specific gene expression by *in situ* hybridization. *Methods in Enzymology*, **152**, 649–61.

Argyris, T.S. (1976). Kinetics of epidermal production during epidermal regeneration following abrasion in mice. *American Journal of Pathology*, **83**, 329–40.

Barrandon, Y. & Green, H. (1985). Cell size as a determinant of the clone forming ability of human keratinocytes. *Proceedings of the National Academy of Sciences, USA*, **82**, 5390–5394.

Barrandon, Y. & Green, H. (1987). Three clonal types of keratinocyte with different capacities for multiplication. *Proceedings of the National Academy of Sciences, USA*, **84**, 2302–6.

Bell, E., Ehrlich, H.P., Buttle, D.J. & Nakatsuji, T. (1981). Living tissue formed *in vitro* and accepted as skin-equivalent tissue of full thickness. *Science*, **211**, 1052–4.

Bell, E., Ivarsson, B. & Merrill, C. (1979). Production of a tissue-like structure by contraction of collagen lattices by human fibroblasts of different proliferative potential *in vitro*. *Proceedings of the National Academy of Sciences, USA*, **76**, 1274–8.

Bendayan, M. & Zollinger, M. (1983). Ultrastructural localization of antigenic sites on osmium-fixed tissues applying the protein A-gold technique. *Journal of Histochemistry and Cytochemistry*, **31**, 101–9.

Bertolino, A.P., Gibbs, P.E.M. & Freedberg, I.M. (1982). *In vitro* biosynthesis of mouse hair keratins under the direction of follicular RNA. *Journal of Investigative Dermatology*, **79**, 173–7.

Bickenbach, J.R. (1981). Identification and behaviour of label retaining cells in oral mucosa and skin. *Journal of Dental Research*, **60**, 1620–2.

Bilbo, P.R., Nolte, C.J.M., Oleson, M.A., Mason, V.S. & Parenteau, N.L. (1992). Skin in organotypic culture: the transition form 'culture' phenotype to organotypic phenotype. *Journal of Toxicology – Cutaneous & Ocular Toxicology*, **12**, 183–96.

Bligh, E.G. & Dyer, W.J. (1959). A rapid method of total lipid extraction and purification. *Canadian Journal of Biochemical Physiology*, **37**, 911–17.

Boyce, S., Christianson, D. & Hansbrough, J. (1988). Structure of a collagen-GAG skin substitute optimized for cultured human epidermal keratinocytes. *Journal of Biomaterials Research*, **22**, 939–57.

Boyce, S. & Hansbrough, J.F. (1988). A skin autograft substitute: biological attachment and growth of cultured human keratinocytes on a graftable collagen and chondroitin-6-sulfate substrate. *Surgery*, **103**, 421–31.

Boyce, S., Michel, S., Reichert, U., Shroot, B. & Schmidt, R. (1990). Reconstructed skin from cultured human keratinocytes and fibroblasts on a collagen–glycosaminoglycan biopolymer substrate. *Skin Pharmacology*, **3**, 136–43.

Brennan, J.K., Mansky, J., Roberts, G. & Lichtman, M.A. (1975). Improved methods for reducing calcium and magnesium concentrations in tissue culture medium: application to studies of lymphoblast proliferation *in vitro*. *In vitro*, **11**, 354–60.

Brunner, K.T., Engers, H.D. & Cerottini, J-C. (1976). The ^{51}Cr release assay used for the quantitative measurement of cell-mediated cytolysis in vitro. In *In vitro Methods in Cell-Mediated and Tumor Immunity*. ed. B.R. Bloom & J.R. David, pp. 423–428. New York: Academic Press.

Bullock, G.R. & Petrusz, P. (1982). *Techniques in Immunocytochemistry*, vol. 1, London, New York: Academic Press.

Cairns, J. (1975). Mutational selection and the natural history of cancer. *Nature*, **255**, 197–200.

Capper, S.J. (1993). Immunoassays for growth factors. In *Growth Factors: A Practical Approach*, ed. I.A. McKay & I.M. Leigh, Practical Approach Series. Oxford: Oxford University Press.

Carlemalm, E., Garavito, R.M. & Villiger, W. (1982). Resin development for electron microscopy and an analysis of embedding at low temperature. *Journal of Microscopy*, **125**, 123–43.

Chakravarty, R. & Rice, R.H. (1989). Acylation of keratinocyte transglutaminase by palmitic and myristic acids in the membrane anchorage region. *Journal of Biological Chemistry*, **264**, 625–9.

Chakravarty, R.C., Rong, X. & Rice, R.H. (1990). Phorbol ester-stimulated phosphorylation of keratinocyte transglutaminase in the membrane anchorage region. *Biochemical Journal*, **271**, 25–30.

Chirgwin, J.M., Przybyla, A.E., MacDonald, R.J. & Rutter, W.J. (1979). Isolation of biologically active ribonucleic acid from sources enriched in ribonuclease. *Biochemistry*, **18**, 5294–9.

Christophers, E. (1971). Cellular architecture of the stratum corneum. *Journal of Investigative Dermatology*, **56**, 165–9.

Clarke, D.D., Mycek, M.J., Neidle, A. & Waelsch, H. (1959). The incorporation of amines into protein. *Archives of Biochemistry and Biophysics*, **79**, 338–54.

Clausen, O.P.F., Elgjo, K., Kirkhus, B., Pedersen, S. & Bolund, L. (1983). DNA synthesis in mouse epidermis: S phase cells that remain unlabelled after pulse labelling with DNA precursors progress slowly through S. *Journal of Investigative Dermatology*, **81**, 545–9.

Clonetics Inc. (1989). Organization of tissue culture systems: reaching a happy medium. *Cellular Communications*, Vol. 1, No. 4.

Cohen, R., Zimber, M., Hansbrough, J.F., Fung, Y.C., Debes, J. & Skalak, R. (1992). Tear strength properties of a novel cultured dermal tissue model. 2nd Annual Meeting. The Wound Healing Society, Richmond, VA.

Cooper, M.L., Andree, C., Hansbrough, J.F., Zapata-Sirvent, R.L. & Spielvogel, R.L. (1993). Direct comparison of a cultured composite skin substitute containing human keratinocytes to an epidermal sheet graft containing human keratinocytes on athymic mice. *Journal of Investigative Dermatology*. **101**, 811–19.

Cooper, M.L. & Hansbrough, J.F. (1991). Use of a composite skin graft composed of cultured human keratinocytes and fibroblasts and a collagen-GAG matrix to cover full-thickness wounds on athymic mice. *Surgery*, **109**, 198–207.

Cooper, M.L., Hansbrough, J.F., Spielvogel, R.L., Cohen, R., Bartel, R.L. & Naughton, G. (1991). *In vivo* optimization of a living dermal substitute employing cultured human fibroblasts on a biodegradable polyglycolic acid or polyglactin mesh. *Biomaterials*, **12**, 243–8.

Cotsarelis, G., Sun, T-T. & Lavker, R.M. (1990). Label retaining cells reside in the bulge area of pilosebaceous unit: implications for follicular stem cells hair cycle and skin carcinogenesis. *Cell*, **61**, 1329–37.

Coussons, P.J., Price, N.C., Kelly, S.M., Smith, B. & Sawyer, L. (1992). Factors that govern the specificity of transglutaminase-catalyzed modification of proteins and peptides. *Biochemical Journal*, **282**, 929–30.

D'Anna, F., De Luca, M., Cancedda, R., Zicca, A. & Franzi, A.T. (1988). Elutriation of human keratinocytes and melanocytes from *in vitro* cultured epithelium. *Histochemistry Journal*, **20**, 674–8.

De Luca, M., D'Anna, F., Bondanza, S., Franzi, A.T. & Cancedda, R. (1988). Human epithelial cells induce human melanocyte growth *in vitro* but only skin keratinocytes regulate the proper differentiation in the absence of dermis. *Journal of Cell Biology*, **107**, 1919–26.

Di Marco, E., Marchisio, P.C., Bondanza, S., Franzi, A.T., Cancedda, R. & De Luca, M. (1991). Growth-regulated synthesis and secretion of biologically active nerve growth factor by human keratinocytes. *Journal of Biological Chemistry*, **266**, 21718–22.

Djian, P., Phillips, M. Easley, K. et al. (1993). The involucrin genes of the mouse and the rat: study of their shared repeats. *Mol. Biol. Evol.* In press.

Doran, T.I., Baff, R., Jacobs, P. & Pacia, E. (1991). Characterization of human sebaceous cells *in vitro*. *Journal of Investigative Dermatology*, **96**, 341–8.

Doran, T.I. & Shapiro, S.S. (1990). Retinoid effects on sebocyte proliferation.

Methods in Enzymology, **190**, 334–8.

Dover, R. & Potten, C.S. (1983). Cell cycle kinetics of cultured human epidermal keratinocytes. *Journal of Investigative Dermatology*, **80**, 423–9.

Eady, R.A.J. (1985). Transmission electron microscopy. *Methods in Skin Research*, ed. D. Skerrow, C.J. Skerrow pp. 1–36. Chichester: Wiley.

Eady, R.A.J. & Shimizu, H. (1992). Electron microscopic immunocytochemistry in dermatology. In *Electron Microscopic Immunocytochemistry: Principles and Practice*, ed. J.M. Polak, & J.V. Priestley pp. 207–222. Oxford: Oxford University Press.

Eckert, R.L. & Green, H. (1986). Structure and evolution of the human involucrin gene. *Cell*, **46**, 583–9.

Eckert, R.L., Yaffe, M.B., Crish, J.F., Murthy, S., Rorke, E.A. & Welter, J.F. (1993). Involucrin – structure and role in envelope assembly. *Journal of Investigative Dermatology*, **100**, 613–17.

Elder, J.T., Fisher, G.J., Linquist, P.B. *et al.* (1989). Overexpression of transforming growth factor α in psoriatic epidermis. *Science*, **243**, 811–14.

Ellis, W.J. (1948). Method for obtaining wool roots for histochemical examination. *Nature*, **162**, 957–60.

Etoh, H., Taguchi, Y.H. & Tabachnik, J. (1977). Cytokinetics of regeneration in β irradiated guinea pig epidermis. *Radiation Research*, **71**, 108–18.

Etoh, U., Simon, M. & Green, H. (1986). Involucrin acts as a transglutaminase substrate at multiple sites. *Biochemical and Biophysical Research Communications*, **136**, 51–6.

Feinberg, A.P. & Vogelstein, B. (1983). A technique for radiolabeling DNA restriction endonuclease fragments to high specific activity. *Analytical Biochemistry*, **132**, 6–13.

Fesus, L., Szucs, E.F., Barrett, K.E., Metcalfe, D.D. & Folk, J.E. (1985). Activation of transglutaminase and production of protein-bound γ-glutamylhistamine in stimulated mouse mast cells. *Journal of Biological Chemistry*, **260**, 13771–8.

Fink, M.L., Shao, Y.Y. & Kersh, G.J. (1992). A fluorometric, high-performance liquid chromatographic assay for transglutaminase activity. *Analytical Biochemistry*, **201**, 270–6.

Folk, J.E. & Chung, S.I. (1973). Molecular and catalytic properties of transglutaminases. *Advances in Enzymology*, **38**, 109–91.

Fuchs, E. (1990). Epidermal differentiation: the bare essentials. *Journal of Cell Biology*, **111**, 2807–14.

Garlick, J.A., Katz, A.B., Fenjves, R.S. & Taichman, L.B. (1991). Retrovirus-mediated transduction of cultured epidermal keratinocytes. *Journal of Investigative Dermatology*, **97**, 824–9.

Gerrard, A.J., Hudson, D.L., Brownlee, G.G. & Watt, F.M. (1993). Towards gene therapy for haemophilia B using normal human keratinocytes. *Nature Genetics*, **3**, 180–3.

Gherzi, R., Melioli, G., De Luca, M. *et al.* (1992). 'HepG2/erythroid/brain' type glucose transporter (GLUT1) is highly expressed in human epidermis: keratinocyte differentiation affects GLUT1 levels in reconstituted epidermis. *Journal of Cellular Physiology*, **150**, 463–74.

Gillespie, J.M. (1991). The structural proteins of hair: isolation, characterisation and regulation of biosynthesis. In *Physiology, Biochemistry and Molecular Biology of the Skin*, 2nd. edn, Vol. 1, ed. L.A. Goldsmith, pp. 625–59. New York: Oxford University Press.

Goding, J.W. (1976). Conjugation of antibodies to fluorochromes – modifications to the standard methods. *Journal of Immunology Methods*, **13**, 215–26.

Gorman, J.J. & Folk, J.E. (1980). Structural features of glutamine substrates for human plasma factor XIIIa (activated blood coagulation factor XIII). *Journal of Biological Chemistry*, **255**, 419–27.

Gottlieb, A.B., Chang, C.K., Posnett, D.N., Fanelli, B. & Tam, J.P. (1988). Detection of transforming growth factor α in normal, malignant and hyper-proliferative human keratinocytes. *Journal of Experimental Medicine*, **167**, 670–5.

Green, H. (1977). Terminal differentiation of cultured human epidermal cells. *Cell*, **11**, 405–16.

Green, H. (1980). The keratinocyte as differentiated cell type. In *The Harvey Lectures, Series 74* New York, Academic Press, pp. 101–139.

Green, H. & Djian, P. (1992). Consecutive actions of different gene-altering mechanisms in the evolution of involucrin. *Molecular Biology and Evolution*, **99**, 977–1017.

Green, H., Kehinde, O. & Thomas, J. (1979). Growth of cultured human epidermal cells into multiple epithelia suitable for grafting. *Proceedings of the National Academy of Sciences, USA*, **76**, 5665–8.

Green, M.R., Clay, C.S., Gibson, W.T. *et al.* (1986). Rapid isolation in large numbers of intact, viable, individual hair follicles from skin: biochemical and ultrastructural characterization. *Journal of Investigative Dermatology*, **87**, 768–70.

Greenberg, C.S., Birchbichler, P.J. & Rice, R.H. (1991). Transglutaminases: multifunctional cross-linking enzymes that stabilize tissues. *FASEB Journal*, **5**, 3071–7.

Grossman, R.M., Krueger, J., Yourish, D. *et al.* (1989). Interleukin 6 is expressed in high levels in psoriatic skin and stimulates proliferation of cultured human keratinocytes. *Proceedings of the National Academy of Sciences, USA*, **86**, 6367–71.

Guy, R., Ridden, C., Barth, J. & Kealey, T. (1993). Isolation and maintenance of the human pilosebaceous duct: 13 *cis*-retinoic acid acts directly on the duct *in vitro*. *British Journal of Dermatology*. **128**. 242–8.

Hachisuka, H., Nomura, H., Mori, O. *et al.* (1990). Alterations in membrane fluidity during keratinocyte differentiation measured by fluorescence polariza-tion. *Cell and Tissue Research*, **260**, 207–10.

Haftek, M., Chignol, M-C. & Thivolet, J. (1989). Quantitative studies of keratin expression with the post-embedding immunogold labelling method. *Journal of Histochemistry and Cytochemistry*, **37**, 735–41.

Hansbrough, J.F., Boyce, S.T., Cooper, M.L. & Foreman, T.J. (1989). Burn wound closure with cultured autologous keratinocytes and fibroblasts attached to a collagen-glycosaminoglycan substrate. *Journal of the American Medical Association*, **262**, 2125–30.

Hansbrough, J.F., Cooper, M.L., Cohen, R. *et al.* (1992). Evaluation of a biodegradable matrix containing cultured human fibroblasts as a dermal replacement beneath meshed skin grafts on athymic mice. *Surgery*, **111**, 438–46.

Hansbrough, J.F., Dore, C. & Hansbrough, W.H. (1992). Clinical trials of a living dermal tissue replacement placed beneath meshed split-thickness skin graft on excised burn wounds. *Journal of Burn Care Rehabilitation*, **13**, 519–29.

Hansbrough, J.F., Moragn, J.L., Greenleaf, G.E. & Bartel, R. (1993). Composite grafts of human keratinocytes grown on a polyglactin mesh-cultured fibroblast dermal substitute function as a bilayer skin replacement in full-thickness wounds on athymic mice. *Journal of Burn Care Rahabitation*, **14**, 485–94..

Harlow, E. & Lane, D.P. (1988). *Antibodies: A Laboratory Manual*. Cold Spring

Harbor, New York: Cold Spring Harbor Laboratory Press.

Heid, H.W., Werner, E. & Franke, W.W. (1986). The complement of native α-keratin polypeptides of hair-forming cells: a subset of eight polypeptides that differ from epithelial cytokeratins. *Differentiation*, **32**, 101–19.

Heimer, C.V. & Taylor, C.E.D. (1974). Improved mountant for immuno-fluorescence preparations. *Journal of Clinical Pathology*, **27**, 254–6.

Hennings, H., Michael, D., Cheng, M.D., Steinert, P., Holbrook, K. & Yuspa, S.H. (1980).. Calcium regulation of growth and differentiation of mouse epidermal cells in culture. *Cell*, **19**, 245–54.

Hennings, H. & Holbrook, K.A. (1983). Calcium regulation of cell–cell contact and differentiation of epidermal cells in culture. An ultrastructural study. *Experimental Cell Research*, **143**, 127–42.

Hennings, H., Michael, D., Lichti, U. & Yuspa, S.H. (1987). Response of carcinogen-altered mouse epidermal cells to phorbol ester tumor promoters and calcium. *Journal of Investigative Dermatology*, **88**, 60–5.

Hennings, H., Robinson, V.A., Michael, D.M., Pettit, G.R., Jung, R. & Yuspa, S.H. (1990). Development of an *in vitro* analogue of initiated mouse epidermis to study tumor promoters and antipromoters. *Cancer Research*, **50**, 4794–800.

Horisberger, M. & Rosset, J. (1977). Colloidal gold, a useful marker for transmission and scanning electron microscopy. *Journal of Histochemistry and Cytochemistry*, **25**, 295–305.

Hsu, S.M., Raine, L. & Fanger, H. (1981). Use of avidin–biotin–peroxidase complex (ABC) in immunoperoxidase techniques. *Journal of Histochemistry and Cytochemistry*, **29**, 577–80.

Hudson, D.L., Weiland, K.L., Dooley, T.P., Simon, M. & Watt, F.M. (1992). Characterisation of eight monoclonal antibodies to involucrin. *Hybridoma*, **11**, 367–79.

Hume, W.J. & Potten, C.S. (1976). The ordered columnar structure of mouse filiform papillae. *Journal of Cell Science*, **22**, 149–60.

Hume, W.J. & Potten, C.S. (1982). A long-lived thymidine pool in epithelial stem cells. *Cell Tissue Kinetics*, **15**, 49–58.

Ishida-Yamamoto, A., McGrath, J.A., Chapman, S.J., Leigh, I.M., Lane, E.B. & Eady, R.A.J. (1991). Epidermolysis bullosa simplex (Dowling–Meara type) is a genetic disease characterized by an abnormal keratin filament network involving keratins K5 and K14. *Journal of Investigative Dermatology*, **97**, 959–68.

Jahoda, C. (1992). Induction of follicle formation and hair-growth by vibrissa dermal papillae implanted into rat ear wounds – vibrissa-type fibres are specified. *Development*, **115**(4), 1103.

Johnson, E.W., Meunier, S.F., Roy, C.J. & Parenteau, N.L. (1992). Serial cultivation of normal human keratinocytes: a defined system for studying the regulation of growth and differentiation. *In vitro Cell Developmental Biology*, **28A**, 429–35.

Johnson, K.S., Harrison, G.B.L., Lighttowlers, M.W. *et al.* (1989). Vaccination against ovine cysticercosis using a defined recombinant antigen. *Nature*, **338**, 585–7.

Jones, P.H. & Watt, F.M. (1993). Separation of human epidermal stem cells from transit amplifying cells on the basis of differences in integrin function and expression. *Cell*, **73**, 713–24.

Kangesu, T., Navsaria, H.A., Manek, S. *et al.* (1993). A porcine model using skin graft chambers for studies on cultured keratinocytes. *British Journal of Plastic*

Surgery, **46**, 393.

Kealey, T., Lee, C.M., Thody, A.J. & Coaker, T. (1986). The isolation of human sebaceous glands and apocrine sweat glands by shearing. *British Journal of Dermatology*, **114**, 181–8.

Khandke, L., Krane, J.F., Ashinoff, R. *et al.* (1991). Cyclosporine in psoriasis treatment. Inhibition of keratinocyte cell-cycle progression in G1 independent of effects on transforming growth factor alpha/epidermal growth factor receptor pathways. *Archives of Dermatology*, **127**, 1172–9.

Kilkenny, A.E., Morgan, D., Spangler, E.F. & Yuspa, S.H. (1985). Correlation of initiating potency of skin carcinogens with potency to induce resistance to terminal differentiation in cultured mouse keratinocytes. *Cancer Research*, **45**, 2219–25.

Kim, H.C., Lewis, M.S., Gorman, J.J. *et al.* (1990). Protransglutaminase from guinea pig skin. Isolation and partial characterization. *Journal of Biological Chemistry*, **265**, 21971–8.

Klein, J.D., Guzman, E. & Kuehn, G.D. (1992). Purification and partial characterization of transglutaminase from *Physarum polycephalum*. *Journal of Bacteriology*, **174**, 2599–605.

Kopan, R. & Fuchs, E. (1989). The use of retinoic acid to probe the relation between hyperproliferation-associated keratins and cell proliferation in normal malignant keratinocytes. *Journal of Cell Biology*, **109**, 295–307.

Krebs, H.A. & Henseleit, K. (1932). Untersuchungen uber die harnstoffbilding im tierkorper. *Hoppe-Seyler's Zeitschrift Physiologische Chemie*, **210**, 143–8.

Kulesz-Martin, M.F., Koehler, B., Hennings, H. & Yuspa, S.H. (1980). Quantitative assay for carcinogen altered differentiation in mouse epidermal cells. *Carcinogenesis*, **1**, 995–1006.

Kvedar, J.C., Pion, I.A., Bilodeau, E.B., Baden, H.P. & Greco, M.A. (1992). Detection of substrates of keratinocyte transglutaminase *in vitro* and *in vivo* using a monoclonal antibody to dansylcadaverine. *Biochemistry*, **31**, 49–56.

Lajtha, L.G. (1963). On the concept of the cell cycle. *Journal of Cell Comparative Physiology*, **62** suppl. 1, 143–5.

Laemmli, U.K. (1970). Cleavage of structural proteins during the assembly of the head of the bacteriophage T4. *Nature*, **277**, 680–5.

Landeen, L.K., Zeigler, F.C., Halberstadt, C., Cohen, R. & Slivka, S.R. (1992). Characterization of a human dermal replacement. *Wounds*, **4**, 167–75.

Lane, D.P. & Lane, E.B. (1981). A rapid antibody assay system for screening hybridoma cultures. *Journal of Immunology Methods*, **47**, 303–7.

Lane, E.B. & Alexander, C.M. (1990). Use of keratin antibodies in tumor diagnosis. *Seminars Cancer Biology*, **1**, 165–79.

Lane, E.B., Wilson, C.A., Hughes, B.R. & Leigh, I.M. (1991). Stem cells in hair follicles: cytoskeletal studies. *Annals of the New York Academy of Science*, **642**, 197–213.

Lavker, R.M. & Sun, T-T. (1982). Heterogeneity in epidermal basal keratinocytes: morphological and functional considerations. *Science*, **215**, 1239–41.

Leblond, C.P., Greulich, R.C. & Periera, J.P.M. (1964). Relationship of cell formation and cell migration in the renewal of stratified squamous epithelia. In *Advance Biology of Skin* Vol 5, p 39. Ed. W. Montagna & R.E. Billingham, NY: Pergamon Press.

Lechner, J.F., Haugen, A.A., McClendon, I.H. & Pettis, E.W. (1982). Clonal growth of normal human bronchial epithelial cells in a serum free medium. *In*

Vitro, **18**, 633–42.

Leigh, I.M., McKay, I., Carver, N., Navsaria, H. & Green, C. (1991). Skin equivalents and cultured skin: from the petri dish to the patient. *Wounds*, **3**, 141–8.

Lenoir, M.C., Bernard, B.A., Ferracin, J., Shroot, B. & Vermorken, A.J.M. (1985). Growth and differentiation of human keratinocytes cultured on eye lens capsules. *Archives of Dermatological Research*, **278**, 120–5.

Levitt, M.L., Gazdar, A.F., Oie, H.K., Schuller, H. & Thacher, S.M. (1990). Cross-linked envelope-related markers for squamous differentiation in human lung cancer cell lines. *Cancer Research*, **50**, 120–8.

Limat, A. & Noser, F.K. (1986). Serial cultivation of single keratinocytes from the outer root sheath of human scalp hair follicles. *Journal of Investigative Dermatology*, **87**, 485–8.

Lindberg, K. & Rheinwald, J.G. (1989). Suprabasal 40 kd keratin (K19) expression as immunohistologic marker of premalignancy in oral epithelium. *American Journal of Pathology*, **134**, 89–98.

Ljunggren, C.A. (1897). Von der Fähigkeit des Hautepithels, ausserhalb des Organismus sein Leben zu behalten mit Berücksichtigung der Transplantation. *Deutsche Zeitschrift für Chirurgie* **47**, 608–15.

Longley, J., Ding, T.G., Cuono, C. *et al.* (1991). Isolation, detection and amplification of mRNA from dermatome strips, epidermal sheets and sorted epidermal cells. *Journal of Investigative Dermatology*, **97**, 974–9.

Lorand, L. (1972). Fibrinoligase: The fibrin-stabilizing factor system of blood plasma. *Annals of the New York Academy of Sciences*, **202**, 6–30.

Lorand, L. & Conrad, S.M. (1984). Transglutaminases. *Molecular and Cellular Biochemistry*, **58**, 9–35.

Lowry, O.H., Rosebrough, N.J., Fall, A.L. & Randall, R.J. (1951). Protein measurement with the folin phenol reagent. *Journal of Biological Chemistry*, **193**, 265–75.

McGrath, J.A., Ishida-Yamamoto, A., Shimizu, H., Fine, J.D. & Eady R.A.J. (1994). Immunoelectron microscopy of skin basement membrane zone antigens: a pre-embedding method using 1-nm immunogold with silver enhancement. *Acta Dermatovenereologica, in press.*

McKay, I.A. (1993). Types of growth factor activity; detection and characterisation of new growth factor activities. In *Growth Factors: A Practical Approach*, ed. I. A. McKay & I. M. Leigh, Practical Approach Series. Oxford: Oxford University Press.

McKay, I.A. & Leigh, I.M. (1991). Epidermal cytokines and their roles in cutaneous wound healing. *British Journal of Dermatology*, **124**, 513–18.

Mackenzie, I.C., Zimmerman, K. & Peterson, L. (1981). The pattern of cellular organization of human epidermis. *Journal of Investigative Dermatology*, **76**, 459–61.

McLean, I. & Nakane, P.K. (1974). Periodate-lysine-paraformaldehyde fixative, a new fixative for immunoelectron microscopy. *Journal of Histochemistry and Cytochemistry*, **22**, 1077–83.

Magee, A.I., Lytton, N.A. & Watt, F.M. (1987). Calcium-induced changes in cytoskeleton and motility of cultured human keratinocytes. *Experimental Cell Research*, **172**, 43–53.

Marshall, R.C. & Gillespie, J.M. (1977). The keratin proteins of wool, horn and hoof from sheep. *Australian Journal of Biological Sciences*, **30**, 389–400.

Marshall, R.C. & Gillespie, J.M. (1982). Comparison of samples of human hair by two dimensional electrophoresis. *Journal of the Forensic Science Society*, **22**, 377–85.

Matoltsy, A.G. (1960). Epidermal cells in culture. *International Reviews in Cytology* **10**, 315–51.

Means, G.E. & Feeney, R.E. (1968). Reductive alkylation of amino groups in proteins. *Biochemistry*, **7**, 2192–201.

Mehta, K., Rao, U.R., Vickery, A.C. & Fesus, L. (1992) Identification of a novel transglutaminase from *Brugia malayi* filarial parasite and its role in growth and development. *Molecular and Biological Parasitology*, **53**, 1–16.

Merighi, A. (1992). Post-embedding electron-microscopic immunocytochemistry. In *Electron Microscopic Immunocytochemistry Principles and Practice*. ed. J.M. Polak, & J.V. Priestley, pp. 51–87. Oxford: Oxford University Press.

Messenger, A.G. (1984). The culture of dermal papilla cells from human hair follicles. *British Journal of Dermatology*, **110**, 685–9.

Michel, S., Courseaux, A., Miquel, C. *et al.* (1991). Determination of retinoid activity by an enzyme-linked immunosorbent assay. *Analytical Biochemistry*, **192**, 232–6.

Michel, S., Schmidt, R., Robinson, S.M., Shroot, B. & Reichert, U. (1987). Identification and subcellular distribution of cornified envelope precursor proteins in the transformed human keratinocyte line SV-K14. *Journal of Investigative Dermatology*, **88**, 301–5.

Michel, S., Schmidt, R., Schroot, B. & Reichert, U. (1988). Morphological and biochemical characterization of the cornified envelopes from human epidermal keratinocytes of different origin. *Journal of Investigative Dermatology*, **91**, 11–15.

Miller, A.D. & Rosman, G.J. (1989). Improved retroviral vectors for gene transfer and expression. *Biotechniques*, **7**, 980–90.

Mischke, D. & Wilde, G. (1987). Polymorphic keratins in human epidermis. *Journal of Investigative Dermatology*, **88**, 191–7.

Mitra, R.S. & Nickoloff, B.J. (1992). Epidermal growth factor and transforming growth factor-alpha decrease gamma interferon receptors and induction of intercellular adhesion molecule (ICAM-1) on cultured keratinocytes. *Journal of Cell Physiology*, **150**, 264–8.

Moll, R., Schiller, D.L. & Franke, W.W. (1990). Identification of protein IT of the intestinal cytoskeleton as a novel type I cytokeratin with unusual properties and expression patterns. *Journal of Cell Biology*, **111**, 567–80.

Morgan, D., Welty, D., Greenhalgh, D., Hennings, H. & Yuspa, S.H. (1992). Development of an in vitro model to study carcinogen-induced neoplastic progression of initiated mouse epidermal cells. *Cancer Research*, **52**, 3145–56.

Morgenstern, J.P. & Land, H. (1991). Choice and manipulation of retroviral vectors. In *Methods in Molecular Biology, Vol 7: Gene Transfer and Expression Protocols*. ed E.J. Murray. Clifton, NJ: Humana Press Inc.

Morris, R.J., Fischer, S.M. & Slaga, T.J. (1985). Evidence that the central and peripherally located cells in the murine epidermal proliferative unit are two distinct populations. *Journal of Investigative Dermatology*, **84**, 277–81.

Morris, R.J., Fischer, S.M. & Slaga, T.J. (1986). Evidence that a slowly cycling subpopulation of adult murine epidermal cells retain carcinogen. *Cancer Research*, **46**, 3061–6.

Morris, R.J., Haunes, A.C., Fischer, S.M. & Slaga, T.J. (1991). Concomitant proliferation and formation of a stratified epithelial sheet by explant outgrowth of epidermal keratinocytes from adult mice. *In Vitro Cell Developmental Biology*, **27A**, 886–95.

Morrison, A.I., Keeble, S. & Watt, F.M. (1988). The peanut lectin-binding glycoproteins of human epidermal keratinocytes. *Experimental Cell Research*, **177**, 247–56.

Nagae, S., Lichti, U., De Luca, L.M. & Yuspa, S.H. (1987). Effects of retinoic acid on cornified envelope formation: Difference between spontaneous envelope formation *in vivo* or *in vitro* and expression of envelope competence. *Journal of Investigative Dermatology*, **89**, 51–8.

Newman, G.R. & Hobot, J.A. (1987). Modern acrylics for post-embedding immunostaining techniques. *Journal of Histochemistry and Cytochemistry*, **35**, 971–81.

Nicholson, L.J. & Watt, F.M. (1991). Decreased expression of fibronectin and the $\alpha_5 \beta_1$ integrin during terminal differentiation of human keratinocytes. *Journal of Cell Science*, **98**, 225–32.

Nickoloff, B.J. (1988). The role of interferon-γ in cutaneous trafficking of lymphocytes with emphasis on molecular and cellular adhesion events. *Archives of Dermatology*, **124**, 1835–43.

Nickoloff, B.J. & Mitra, R.S. (1989). Inhibition of ^{125}I-epidermal growth factor binding to cultured keratinocytes by antiproliferative molecules gamma interferon, cyclosporin A, and transforming growth factor-beta. *Journal of Investigative Dermatology*, **93**, 799–803.

Nickoloff, B.J. & Mitra, R.S. (1992). Intraepidermal psoriatic cytokine network involves gamma interferon, transforming growth factor-alpha, and their cell surface receptors: dysregulation rather than deficiency. *Journal of Investigative Dermatology*, (letter). In press.

Nickoloff, B.J., Mitra, R.S., Elder, J.T., Fisher, G.J. & Voorhees, J.J. (1989). Decreased growth inhibition by recombinant gamma interferon is associated with increased transforming growth factor-α production in keratinocytes cultured from psoriatic lesions. *British Journal of Dermatology*, **121**, 161–74.

Noser, F.K. & Limat, A. (1987). Organotypic culture of outer root cells from human hair follicles using a new culture device. *In vitro Cell Developmental Biology*, **23**, 541–5.

O'Connor, N.E., Mulliken, J.B., Banks-Schlegel, S. *et al.* (1981). Grafting of burns with cultured epithelium prepared from autologous epidermal cells. *Lancet*, **i**, 75–8.

Oettinger, H.F., Pasqualini, R. & Bernfield, M. (1992). Recombinant peptides as immunogens: a comparison of protocols for antisera production using the pGEX system. *BioTechniques*, **12**, 544–9.

O'Farrell, P.Z., Goodman, H.M. & O'Farrell, P.H. (1977). High resolution two-dimensional electrophoresis of basic as well as acidic proteins. *Cell*, **12**, 1133–42.

O'Guin, W.M., Schermer, A., Lynch, M. & Sun, T.-T. (1990). Differentiation-specific expression of keratin pairs. In *Cellular and Molecular Biology of Intermediate Filaments*, ed. R.D. Goldman, & P.M. Steinert, pp. 301–334. New York: Plenum Press.

O'Keefe, E.J., Briggaman, R.A. & Herman, B. (1987). Calcium-induced assembly of adherens junctions in keratinocytes. *Journal of Cell Biology*, **105**, 807–17.

Ormerod, M.G.E., ed. (1990). *Flow Cytometry – A Practical Approach*. Oxford: IRL Press.

Parameswaran, K.N., Velasco, P.T., Wilson, J. & Lorand, L. (1990). Labeling of ε-lysine cross-linking sites in proteins withe peptide substrate of factors XIIIIa and transglutaminase. *Proceedings of the National Academy of Sciences, USA,* **87,** 8472–5.

Parenteau, N.L., Eckert, R.L. & Rice, R.H. (1987). Primate involucrins: Antigenic relatedness and multiple forms. *Proceedings of the National Academy of Sciences, USA,* **84,** 7571–5.

Parenteau, N.L., Nolte, C.J.M., Bilbo, P. *et al.* (1991). Epidermis generated *in vitro:* practical considerations and applications. *Journal of Cellular Biochemistry,* **45,** 245–51.

Pavlovitch, J.H., Rizk-Rabin, M., Gervaise, M., Metezeau, P. & Grunwald, D. (1989). Cell subpopulations within proliferative and differentiating compartments of epidermis. *American Journal of Physiology and Cell Physiology,* **25,** C986–C997.

Pavlovitch, J.H., Rizk-Rabin, M., Jaffray, P., Hoehn, H. & Poot, M. (1991). Characteristics of homogeneously small keratinocytes from newborn rat skin: possible epidermal stem cells. *American Journal of Physiology and Cell Physiology,* **261,** C964–72.

Peehl, D.M. & Ham, R.G. (1980). Clonal growth of human keratinocytes with small amounts of dialyzed serum. *In Vitro,* **16,** 526–38.

Pei, X.F., Gorman, P.A. & Watt, F.M. (1991). Two strains of human keratinocytes transfected with HPV 16 DNA: comparison with the normal parental cells. *Carcinogenesis,* **12,** 277–84.

Philpott, M.P., Green, M.R. & Kealey, T. (1990). Human hair growth *in vitro. Journal of Cell Science,* **97,** 463–71.

Piacentini, M., Martinet, N., Beninati, S. & Folk, J.E. (1988). Free and protein-conjugated polyamines in mouse epidermal cells. Effect of high calcium and retinoic acid. *Journal of Biological Chemistry,* **263,** 3790–4.

Polak, J.M. & Priestley, J.V. (eds). (1992). *Electronmicroscopic immunocytochemistry. Principles and Practice.* Oxford: Oxford University Press.

Polak, J.M. & Van Noorden, S. (1987). *An Introduction to Immunocytochemistry: Current Techniques and Problems.* Oxford: Oxford University Press.

Potten, C.S. (1974). The epidermal proliferative unit: the possible role of the central basal cell. *Cell Tissue Kinetics,* **7,** 77–88.

Potten, C.S. & Allen, T.D. (1975). The fine structure and cell kinetics of mouse epidermis after wounding. *Journal of Cell Science,* 413–37.

Potten, C.S. (1977). Extreme sensitivity of some intestinal crypt cells to X and gamma irradiation. *Nature,* **269,** 518–21.

Potten, C.S. (1980). Stem cells in small intestinal crypts. In *Cell Proliferation in the Gastrointestinal Tract,* ed. D.R. Appleton, J.P. Sunter, & A.J. Watson, Tunbridge Wells: Pitman Medical.

Potten, C.S. (1981). Cell replacement in epidermis (keratopoiesis) via discrete units of proliferation. *International Review in Cytology,* **69,** 271–318.

Potten, C.S. (1992). The significance of spontaneous and induced apoptosis in the gastrointestinal tract of mice. *Cancer and Metastasis Review,* **11,** 179–95.

Potten, C.S., Al-Bawari, S.E., Hume, W.J. & Searle, J. (1977). Circadian rhythms of presumptive stem cells in three different epithelia of the mouse. *Cell Tissue Kinetics,* **10,** 557–68.

Potten, C.S. & Hendry, J.H. (1973). Clonogenic cells and stem cells in epidermis. *International Journal of Radiation Biology,* **5,** 537–40.

Potten, C.S., Hume, W.J., Reid, P. & Cairns, J. (1978). The segregation of DNA in epithelial stem cells. *Cell*, **15**, 899–906.

Price, J. (1987). Retroviruses and the study of cell lineage. *Development*, **101**, 409–19.

Pruss, R.M., Mirsky, R., Raff, M.C., Thorpe, R., Dowding, A.J. & Anderton, B.H. (1981). All classes of intermediate filaments share a common antigenic determinant defined by a monoclonal antibody. *Cell*, **27**, 419–28.

Rao, J. & Otto, W.R. (1992). Fluorimetric DNA assay for cell growth estimation. *Analytical Biochemistry*, **207**, 186–92.

Regauer, S. & Compton, C. (1990*a*). Cultured porcine epithelial grafts: An improved method. *Investigative Dermatology*, **94**, 230–4.

Regauer, S. & Compton, C.C. (1990*b*). Cultured keratinocyte sheets enhance spontaneous re-epithelialization in a dermal explant model of partial-thickness wound healing. *Journal of Investigative Dermatology*, **95**, 341–6.

Regnier, M., Asselineau, D. & Lenoir, M.C. (1990). Human epidermis reconstructed on dermal substrates *in vitro*: an alternative to animals in skin pharmacology. *Skin Pharmacology*, **3**, 70–85.

Reichert, U., Michel, S. & Schmidt, R. (1993). The cornified envelope: a key structure of terminally differentiating keratinocytes. In *Molecular Biology of the Skin*, Vol. 1, ed. M. Blumenberg & M. Darmon. pp. 107–150. New York: Academic Press.

Rentrop, M., Knapp, B., Winter, H. & Schweizer, J. (1986). Aminoalkylsilane-treated glass slides as support for *in situ* hybridization of keratin cDNAs to frozen tissue sections under varying fixation and pretreatment conditions. *Histochemical Journal*, **18**, 271–6.

Reynolds, A.J. & Jahoda, C.A.B. (1991). Hair follicle stem cells? A distinctive germinative epidermal cell population is activated *in vitro* by the presence of hair follicle dermal papilla cells. *Journal of Cell Science*, **99**, 373–85.

Reynolds, A.J. & Jahoda, C.A.B. (1992). Cultured dermal papilla cells induce, follicle formation and hair-growth by transdifferentiation of an adult epidermis. *Development*, **115**(2), 587–93.

Rheinwald, J.G. (1980). Serial cultivation of normal human keratinocytes. *Methods in Cell Biology*, **21A**, 229–54.

Rheinwald, J.G. (1989). Methods for clonal growth and serial cultivation of normal human epidermal keratinocytes and mesothelial cells. In *Cell Growth and Division. A Practical Approach*, ed. R. Baserga, pp. 81–94. Oxford: IRL Press.

Rheinwald, J.G. & Green, H. (1975). Serial cultivation of strains of human epidermal keratinocytes: the formation of keratinizing colonies from single cells. *Cell*, **6**, 331–43.

Rheinwald, J.G. & Green, H. (1977). Epidermal growth factor and the multiplication of cultured human epidermal keratinocytes. *Nature*, **265**, 421–4.

Rice, R.H. & Cline, P.R. (1984). Opposing effects of 2,3,7,8-tetrachlorodibenzo-*p*-dioxin and hydrocortisone on growth and differentiation of cultured malignant human keratinocytes. *Carcinogenesis*, **5**, 367–71.

Rice, R.H. & Green, H. (1978). Relationship of protein synthesis and transglut-aminase activity to formation of the cross-linked envelope during terminal differentiation of the cultured human epidermal keratinocyte. *Journal of Cell Biology*, **76**, 705–11.

Rice, R.H. & Green, H. (1979). Presence in human epidermal cells of a soluble protein precursor of the cross-linked envelope: Activation of the cross-linking process by calcium ions. *Cell*, **18**, 681–94.

Rice, R.H. & Thacher, S.M. (1986). Involucrin: a constituent of cross-linked envelopes and marker of squamous maturation. In *Biology of the Integument, vol. 2, Vertebrates*, ed. J. Bereiter-Hahn, A.G. Matoltsy & K.S. Richards, pp. 754–61. Berlin Heidelberg: Springer-Verlag.

Rice, R.H., Chakravarty, R., Chen, J., O'Callahan, W. & Rubin, A.L. (1988). Keratinocyte transglutaminase: regulation and release. *Advances in Experimental Biology and Medicine*, **231**, 51–61.

Rice, R.H., Rong, X. & Chakravarty, R. (1988). Suppression of keratinocyte differentiation in SCC-9 human squamous carcinoma cells by benzo(a)pyrene, 12-O-tetradecanoylphorbol-13-acetate and hydroxyurea. *Carcinogenesis*, **9**, 1885–90.

Rice, R.H., Chakravarty, R., Rong, X. & Rubin, A.L. (1989). Cross-linked envelopes: Keratinocyte transglutaminase. In *Symposium on the Biology of Wool and Hair*, ed. G.E. Rogers, P.J. Reis, K.A. Ward & R.C. Marshall, pp. 389–401. U.K.: Chapman and Hall Ltd.

Richardson, K.C., Jarett, L. & Finke, E.H. (1960). Embedding in epoxy resins for ultrathin sectioning in electron microscopy. *Stain Technology*, **35**, 313–23.

Ridden, J., Ferguson, D. & Kealey, T. (1990). Organ maintenance of human sebaceous glands: *in vitro* effects of 13-*cis* retinoic acid and testosterone. *Journal of Cell Science*, **95**, 125–36.

Rogers, G.E. (1959*a*). Newer findings on the enzymes and proteins of hair follicles. *Annals of the New York Academy of Sciences*, **83**, 408–28.

Rogers, G.E. (1959*b*). Electron microscopy of wool. *Journal of Ultrastructure Research*, **2**, 309–30.

Roop, D.R., Huitfeldt, H., Kilkenny, A. & Yuspa, S.H. (1987). Regulated expression of differentiation-associated keratins in cultured epidermal cells detected by monospecific antibodies to unique peptides of mouse epidermal keratins. *Differentiation*, **35**, 143–50.

Rose, J.K., Buonocore, L. & Whitt, M.A. (1991). A new cationic liposome reagent mediating nearly quantitative transfection of animal cells. *Biotechniques*, **10**, 520–5.

Roth, J. (1984). The protein A-gold technique for antigen localization in tissue sections by light and electron microscopy. In *Immunolabelling for Electron Microscopy*, ed. J.M. Polak & I.M. Varndale pp. 113–122. Amsterdam: Elsevier.

Rubin, A.L., Parenteau, N.L. & Rice, R.H. (1989). Coordination of keratinocyte programming in human SCC-13 squamous carcinoma and normal epidermal cells. *Journal of Cellular Physiology*, **138**, 208–14.

Rubin, A.L. & Rice, R.H. (1986). Differential regulation by retinoic acid and calcium of transglutaminases in cultured neoplastic and normal human keratinocytes. *Cancer Research*, **46**, 2356–61.

Rupniak, H.T., Rowlatt, C., Lane, E.B. *et al.* (1985). Characteristics of four new human cell lines derived from squamous cell carcinomas of the head and neck. *Journal of the National Cancer Institute*, **75**, 621–35.

Sakai, L.Y., Keene, D.R., Morris, N.P. & Burgeson, R.E. (1986). Type VII collagen is a major structural component of anchoring fibrils. *Journal of Cell Biology*, **103**, 1577–86.

Sambrook, J., Fritsch, E.F. & Maniatis, T. (1989). *Molecular Cloning: A Laboratory Manual.* Cold Spring Harbor, New York: Cold Spring Harbor Laboratory Press.

Schmidt, G.H., Blount, M.A. & Ponder, B.A. (1987). Immunochemical demonstration of the clonal organization of chimaeric mouse epidermis. *Development*, **100**, 535–41.

Scofield, R. (1978). The relationship between the spleen colony-forming cell and the haemopoietic stem cell: a hypothesis. *Blood Cells*, **4**, 7–25.

Sheibani, N., Rhim, J.S. & Allen-Hoffmann, B.L. (1991). Malignant HPV type 16-transformed human keratinocytes exhibit altered expression of extracellular matrix glycoproteins. *Cancer Research*, **51**, 5967–75.

Shimizu, H., Ishida-Yamamoto, A. & Eady, R.A.J. (1992). The use of silver-enhanced 1-nm gold probes for light and electron microscopic localization of intra- and extracellular antigens in skin. *Journal of Histochemistry and Cytochemistry*, **40**, 883–8.

Shimizu, H., McDonald, J.N., Gunner, D.B. *et al.* (1990). Epidermolysis bullosa acquisita antigen and the carboxy terminus of type VII collagen have a common immunolocalization to anchoring fibrils and lamina densa of basement membrane. *British Journal of Dermatology*, **122**, 577–85.

Shimizu, H., McDonald, J.N., Kennedy, A.R. & Eady, R.A.J. (1989). Demonstration of intra- and extracellular localization of bullous pemphigoid antigen using cryofixation and freeze substitute for post-embedding immunoelectron microscopy. *Archives of Dermatological Research*, **281**, 443–8.

Signorini, M., Beninati, S. & Bergamini, C.M. (1991). Identification of transglutaminase activity in the leaves of silver beet (*Beta vulgaris L.*). *Journal of Plant Physiology*, **137**, 547–52.

Simon, M. & Green, H. (1985). Enzymatic cross-linking of involucrin and other proteins by keratinocyte particulates *in vitro*. *Cell*, **40**, 677–83.

Simon, M. & Green, H. (1989). Involucrin in the epidermal cells of subprimates. *Journal of Investigative Dermatology*, **92**, 721–4.

Slivka, S.R., Landeen, L., Zimber, M.P. & Bartel, R.L. (1991*a*). Biochemical characterization, barrier function and drug metabolism in an *in vitro* skin model. *Journal of Cell Biology*, **115A**, 1370.

Slivka, S.R., Landeen, L., Zimber, M.P. & Bartel, R.L. (1991*c*). Characterization of a three-dimensional human skin culture model for in vitro percutaneous absorption studies. *Pharmaceutical Research*, **8S**, 143.

Slivka, S.R., Zeigler, F. & Bartel, R.L. (1991*b*). An *in vitro* skin model for the study of keratinocyte responses to irritants. *Journal of Cell Biology*, **115A**, 2072.

Smith, D.B. & Johnson, K.S. (1988). Single step purification of polypeptides expressed in *E. coli* as fusion with glutathione-S-transferase. *Gene*, **67**, 31–40.

Staiano-Coico, L., Higgins, P.J., Darzynkiewicz, Z. *et al.* (1986). Human keratinocyte culture – identification and staging of epidermal cell subpopulations. *Journal of Clinical Investigation*, **77**, 396–404.

Stark, H.-J., Breitkreutz, D., Limat, A., Bowden, P. & Fusenig, N.E. (1987). Keratins of the human hair follicle: 'hyperproliferative' keratins consistently expressed in the outer root sheath cells *in vitro* and *in vivo*. *Differentiation*, **35**, 236–48.

Stasiak, P.C., Purkis, P.E., Leigh, I.M. & Lane, E.B. (1989). Keratin 19: predicted amino acid sequence and broad tissue distribution suggest it evolved from keratinocyte keratins. *Journal of Investigative Dermatology*, **92**, 707–16.

Stöhr, P. (1904). Entwicklungsgeschichte des menschlichen Wollhaares. Anat. Hefte Abt.I. **23**, 1–66.

Sun, T.-T. & Green, H. (1976). Differentiation of the epidermal keratinocyte in cell culture: formation of the cornified envelope. *Cell*, **9**, 511–21.

Tatnall, E.M., Leigh, I.M. & Gibson, J.R. (1987). Comparative toxicity of antimicrobial agents on transformed human keratinocytes. *Journal of Investigative Dermatology*, **89**, 316.

Taylor-Papadimitriou, J., Purkis, P., Lane, E.B., McKay, I.A. & Chang, S.E.

(1982). Effects of SV40 transformation on the cytoskeleton and behavioural properties of human keratinocytes. *Cell Differentiation*, **11**, 169–80.

Teumer, J., Zezulak, K. & Green, H. (1994). Measurement of specific mRNA content of keratinocytes of different sizes in relation to growth and differentiation. In *Keratinocyte Handbook* ed. I.M. Leigh, E.B. Lane & F.M. Watt. Cambridge: Cambridge University Press.

Thacher, S.M., Coe, E.L. & Rice, R.H. (1985). Retinoid suppression of transglutaminase activity and envelope competence in cultured human epidermal carcinoma cells: Hydrocortisone is a potent antagonist of retinyl acetate but not retinoic acid. *Differentiation*, **29**, 82–7.

Thacher, S.M. & Rice, R.H. (1985). Keratinocyte-specific transglutaminase of cultured human epidermal cells: relation to cross-linked envelope formation and terminal differentiation. *Cell*, **40**, 685–95.

Torma, H. & Vahlquist, A. (1990). Vitamin A esterification in human epidermis: a relation to keratinocyte differentiation. *Journal of Investigative Dermatology*, **94**, 132–8.

Toye, B., Zhong, G., Peeling, R. & Brunham, R.C. (1990). Immunologic characterization of a cloned fragment containing the species-specific epitope from the major outer membrane protein of chlamydia trachomatis. *Infection Immunology*, **58**, 3909–13.

Triglia, D., Braa, S.S., Yonan, C. & Naughton, G.K. (1991). *In vitro* toxicity of various classes of test agents using the neutral red assay on a human three-dimensional physiologic skin model. *In Vitro Cellular Developments Biology*, **27A**, 239–44.

Turbitt, M.L., Akhurst, R.J., White, S.I. & MacKie, R.M. (1990). Localisation of elevated transforming growth factor-alpha in psoriatic epidermis. *Journal of Investigative Dermatology*, **95**, 229–32.

Van Zoelen, E.J.J., Delaey, B., Van der Burg, B. & Huylebroeck, D. (1993). Detection of polypeptide growth factors: application of specific bioassays and PCR technology. In *Growth Factors: A Practical Approach*, ed. I.A. McKay & I.M. Leigh, Practical Approach Series. Oxford: Oxford University Press.

Warhol, M.J., Roth, J., Lucocq, J.M., Pinkus, G.S. & Rice, R.H. (1985). Immuno-ultrastructural localization of involucrin in squamous epithelium and cultured keratinocytes. *Journal of Histochemistry and Cytochemistry*, **33**, 141–9.

Watt, F.M. (1984). Selective migration of terminally differentiating cells from the basal layer of cultured human epidermis. *Journal of Cell Biology*, **98**, 16–21.

Watt, F.M. (1989). Terminal differentiation of epidermal keratinocytes. *Current Opinion in Cell Biology*, **1**, 1107–15.

Watt, F.M. (1993). Human epidermal keratinocytes in culture: role of integrins in regulating adhesion and terminal differentiation. In *Molecular Basis of Morphogenesis*, ed. M. Bernfield, pp. 241–54. New York: Wiley-Liss Inc.

Watt, F.M. & Green, H. (1981). Involucrin synthesis is correlated with cell size in human epidermal cultures. *Journal of Cell Biology*, **90**, 738–42.

Watt, F.M. & Green, H. (1982). Stratification and terminal differentiation of cultured epidermal cells. *Nature*, **295**, 434–6.

Watt, F.M. & Hertle, M.D. (1994). Keratinocyte integrins. In *The Keratinocyte Handbook*, ed. I.M. Leigh, E.B. Lane & F.M. Watt, Cambridge: Cambridge University Press.

Watt, F.M. & Jones, P.H. (1992). Changes in cell surface carbohydrate during terminal differentiation of human epidermal keratinocytes. *Biochemical Society*

Transactions, **20**, 285–8.

Watt, F.M., Jordan, P.W. & O'Neill, C.H. (1988). Cell shape controls terminal differentiation of human epidermal keratinocytes. *Proceedings of the National Academy of Sciences USA*, **85**, 5576–80.

Watt, F.M., Kubler, M-D., Hotchin, N.A., Nicholson, L.J. & Adams, J.C. (1993). Regulation of keratinocyte terminal differentiation by integrin-extracellular matrix interactions. *Journal of Cell Science*, **106**, 175–82.

Wells, J. (1982). A simple technique for establishing cultures of epithelial cells. *British Journal of Dermatology*, **197**, 481–2.

Weterings, P.J.J.M., Vermorken, A.J.M. & Bloemendal, H. (1981). A method for culturing human hair follicle cells. *British Journal of Dermatology*, **104**, 1–5.

Weterings, P.J.J.M., Roelofs, H.M.J., Vermorken, A.J.M. & Bloemendal, H. (1983). Serial cultivation of human scalp hair follicle keratinocytes. *Acta Dermatologia Venereologia* (Stockh), **63**, 315–20.

Wilkins, L.M., Watson, S.R., Prosky, S.J., Meunier, S.F. & Parenteau, N. L. (1994). Development of a bilayered living skin construct for clinical applications. *Biotechnology & Bioengineering*, **43**, 747–56.

Wilkinson, D.G. (1992). *In situ Hybridization: A Practical Approach*. Oxford: IRL Press.

Wu, Y-J., Parker, L.M., Binder, N.E. *et al.* (1982). The mesothelial keratins: a new family of cytoskeletal proteins identified in culture mesothelial cells and on-keratinising epithelia. *Cell*, **31**, 693–703.

Xia, L., Zouboulis, C.C., Detmar, M., Mayer-da-Silva, A., Stadler, R. & Orfanos, C.E. (1989). Isolation of human sebaceous glands and cultivation of sebaceous gland-derived cells as an *in vitro* model. *Journal of Investigative Dermatology*, **93**, 315–21.

Yannas, I.V., Burke, J.F., Gordon, P.L., Huang, C. & Rubenstein, R.H. (1980). Design of an artificial skin. II: Control of chemical composition. *Journal of Biomedical Materials Research*, **14**, 107–31.

Yoshizaki, K., Nishimoto, N., Matsumoto, K. *et al.* (1990). Interleukin 6 and expression of its receptor on epidermal keratinocytes. *Cytokine*, **2**, 381–7.

Yuspa, S.H. (1985). Methods for the use of epidermal cell culture to study chemical carcinogenesis. In *Methods in Skin Research*, ed. D. Skerrow, & C. Skerrow, pp. 213–249. Chichester: John Wiley.

Yuspa, S.H. & Harris, C.C. (1974). Altered differentiation of mouse epidermal cells treated with retinyl acetate *in vitro*. *Experimental Cell Research*, **86**, 95–105.

Yuspa, S.H., Kilkenny, A.E., Steinert, P.M. & Roop, D.R. (1989). Expression of murine epidermal differentiation markers is tightly regulated by restricted extracellular calcium concentrations *in vitro*. *Journal of Cell Biology*, **109**, 1207–17.

Zouboulis, C.C., Korge, B., Akamatsu, H., Xia, L., Schiller, S., Gollnick, H. & Orfanos, C.E. (1991). Effects of 13 *cis*-retinoic acid, all *trans*-retinoic acid and Acitrenin on the proliferation, lipid synthesis and keratin expression of cultured human sebocytes *in vitro*. *Journal of Investigative Dermatology*, **96**, 792–7.

Index

Location references given in *italics*
indicate diagrams or tables.

3T3 feeder layer
  in flow cytometry, 107
  human keratinocyte cultures, 5–16
  preparation, 7
  in specific bioassay, IL-6, 94
  survival, low calcium medium, 13
$3T\beta$ culture medium (3T3 cm), 6–7
^{51}Cr labelling, 73–4

α-casein, 161, 162
acceptor substrates, 161–62
acrylamide, 139
adenine, 6
agarose beads, 172, 174
air–liquid interface
  skin equivalent cultures at, *49*, 53;
    barrier development, *54*;
    differentiation, epidermal layer,
    materials and methods, 51–2
allografts
  treatment of severe burns, 63, 65
antibody
  binding, visualization, 128–9
  detection, growth factors, 92
  production: against human
    involucrin, 169; recombinant
    involucrin in, 172, *173*, 174
antibodies
  anti-integrin, keratinocyte labelling,
    107
  to involucrin, 158
  monoclonal, 128
antigenic cross-reactivity, related
  species, 158
antigens, intracellular, labelling, 122–3

antipodal cocultures, 60–1
assays
  enzyme-linked immunosorbent, 92;
    involucrin, 170
  extracellular matrix adhesion, 73–4
  fluorimetric DNA assay, 89–90
  growth factor, 91–96
  inophore-inducible envelope, 163–4
  involucrin immunoassay, 157–9
  keratinocyte proliferation, 75–87
  microfluorescence, *131*
  transglutaminase, 159–62
autoradiography
  darkroom technique, 80–83, 120;
    preparation, 79–81
  human hair follicle, 38–9, 42, 152

B9 hybridoma cell line, IL-6 specific
  bioassay, 93–6
β-casein, 161–62
Balb/C 3T3, 5
basal cells, flow cytometry, 108
bioassay, specific, growth factors, 93
  IL-6, 93–6
burns, keratinocyte sheets for grafting,
  63–5
  application to patient, 65
  preparation of graft, 63, 64–5

calcium
  levels, effects of, 21, 25; low calcium
    culture, 13, 21–3
cell
  adhesion, 73–4
  cultures: 3T3 feeder layer, human
    keratinocytes, 5–16; antipodal
    cocultures, 60–61; collagen gels,
    keratinocytes, 55–60; dermal
    replacement tissue, 68;